Field Methods in Remote Sensing

Field Methods
in Remote Sensing

Roger M. McCoy

THE GUILFORD PRESS
New York London

Printed in the United States of America

This book is printed on acid-free paper.

Last digit is print number: 9   8   7   6   5   4   3   2   1

Library of Congress Cataloging-in-Publication Data

McCoy, Roger M.
  Field methods in remote sensing / Roger M. McCoy.
      p.   cm.
  Includes bibliographical references and index.
    ISBN 1-59385-079-4 (pbk.) — ISBN 1-59385-080-8 (hardcover)
    1. Remote sensing—Field work.   I. Title.
  G70.4.M39 2005
  526—dc22

                                            2004019785

To the memory of Professor M. John Loeffler (University of Colorado), with gratitude for showing me the way

Preface

Practitioners of remote sensing will at some point need to learn how to obtain field data suitable for the various needs of their projects.

Effective field data are best obtained through thoughtful planning, thorough knowledge of valid sampling techniques, accurate location-finding procedures, and reliable field measurements.

Unfortunately for the beginner, few remote sensing research reports provide thorough accounts of the methods that were followed in the field. Instead, they concentrate on laboratory procedures such as data correction and processing. However, the methods of measuring field data have as much influence on the reliability of the final product as do laboratory procedures. Field procedures are important and should always be included in final reports as a service to readers who would like to validate, replicate, or educate.

The purpose of this book on field procedures is to ease the way for the person who has a background in the fundamentals of remote sensing and laboratory methods but little practical knowledge of the field methods that may be needed for remote sensing projects. The readers I am envisioning include the following: students with some background in the fundamentals of remote sensing and image data processing who want to begin a project requiring field work; professionals with government agencies who may have field skills but need guidance applying them appropriately to remote sensing; and teachers who want to supplement a remote sensing course with a project requiring field work but whose field experience is limited or rusty.

The field methods discussed encompass project planning, sampling plans suitable for selecting spectral training sites or accuracy assessment sites, finding locations in the field using a global positioning system, obtaining reflectance spectra from objects in the field, and basic measurement methods for studies of vegetation, soils, and water.

The goals of remote sensing projects cover a wide range. The most fundamental is to produce a map of some selected surface features. Others may be to calibrate sensors with the response of surface features, to validate or evaluate the final product, to model the spectral response of a material and its biophysical characteristics, or to develop or test image data processing techniques. Because the approaches to field work in these various types of studies may be similar to each other, this guidebook will not differentiate field methods on that basis. Examples in the book usually assume that a map is the intended result.

Measurement methods, particularly for vegetation, vary widely. Where the choice of measurement methods is large, the selection of an appropriate method depends on the amount of precision and detail needed for the final product. As objectives become increasingly specialized, the method of measurement may be less widely known, and often its effectiveness may be considered controversial among professionals. The plan for this book is to provide several basic measurement methods for vegetation, soils, and water/snow that might be applied to many types of projects. For situations in which more specialized measurements are needed, I have provided a bibliography, in Appendix 1, on advanced field methods. For example, many studies require field data quantifying vegetation cover density. Chapter 6 provides some methods for obtaining these data, in this case by pacing, in which a measurement is taken at each step, or by the analysis of small areas. Methods for deriving volume or weight of vegetation are also presented. For more advanced or specialized methods, the reader would refer to the bibliography in Appendix 1.

The outcome of field work in remote sensing varies widely depending on project objectives. The simplest result may be an aerial photograph annotated with current ground cover types. Another outcome of field work may include a notebook of field data sheets containing measurements and observations for each sample site. In any case the strategy is to relate field data with image data as accurately and thoroughly as possible. The goal of field work should be that data collected from a variety of field sites are representative of the surface "seen" by the airborne or satellite sensor. The immense difference in scales between the ground and the image makes this an especially challenging task.

The discussions of field procedures in this book are always focused on the needs of remote sensing projects that use sensors in the 400- to 2,500-nm wavelength range, that is, reflective radiation. When thermal infrared or microwave wavelengths are used, many of the field methods described here would still apply, but a number of others would need to be added.

Acknowledgments

Without the steady urging and encouragement of my wife, Sue, this book might never have happened. I deeply appreciate her support. Through her I came to understand why so many authors acknowledge their spouse's contribution.

Much is owed to the dozens of students who learned field methods the hard way—by digging through the literature of many disciplines. Their dogged efforts are reflected in this book. Together we learned to appreciate those remote sensing researchers who bothered to describe their field methods in detail.

Special thanks are due for the critiques of Professors Lallie Scott, Chris Johannsen, and Larry Biehl. Each of them made numerous constructive suggestions and pointed out errors and ambiguities. Other reviewers, who were anonymous, also contributed an enormous effort and provided much valuable input. A great many of their suggestions were gladly incorporated into this book.

Contents

Field Methods in Remote Sensing

1

Problems and Objectives in Remote Sensing Field Work

PROBLEMS ENCOUNTERED IN FIELD WORK

Many published remote sensing project reports have a strong emphasis on image processing techniques with very little detail regarding the methods used for collecting data and information in the field. One may read some reports and wonder whether the researchers even found it necessary to do field work or to use maps, aerial photographs, or other reference materials. As a result, new researchers looking for information on remote sensing field methods often must start from scratch by scouring the literature of related disciplines for guidance. Frequently the result is that the field methods for a remote sensing project are poorly planned and the final product has hidden weaknesses that could have been avoided by careful advance planning.

Since most remote sensing projects require some amount of field work, there should be significant benefits to a systematic approach to planning the field portion of the project. Certainly, the final product will be more reliable and defensible if the field work and the use of reference materials are planned and executed properly. Even weaknesses in the final results can be stated openly if the unavoidable deficiencies in field work and reference materials are known and explained. There is a tendency among researchers to avoid mentioning weaknesses in their methods, even when those shortcomings are beyond their control. Eventually someone, perhaps in a thesis defense, will ask about field methods or ancillary materials used, and any shortcomings will come to light. It is best to avoid this embarrassment by recognizing and dealing with these details in advance!

The following is a summary of components that should be considered in planning the field portion of a remote sensing project. The approach to planning field work in remote sensing consists in identifying pitfalls and problems and selecting appropriate solutions in advance (Joyce, 1978). Also, this guide provides some procedures for avoiding problems in the field and for making appropriate measurements and observations. An extensive bibliography on field methods is found in Appendix 1.

Problem 1: Lack of Clear Objectives for the Project

It seems self-evident that one must have objectives before beginning a project. However, often the objectives are not thought out in sufficient detail. A thorough written statement of objectives will set the agenda for the entire project and will determine which methods should be used at every stage of work. The planning of objectives will depend in large measure on the expected result, or the nature of the final product. Whether the result is a map or a research report describing a biophysical model, preliminary planning is essential. The examples given in this section assume that a map is the final product.

Initial planning provides the foundation for all subsequent steps in the project. A comprehensive statement of objectives should include the following items: (1) location and size of area; (2) scale of final maps, if maps are the final product; (3) proposed accuracy of the final result; (4) purpose and end users of the final product (i.e., who will use the final maps or models, and how will they be used); (5) anticipated legend of the final map (initially this is a rational legend, based on what one hopes to show on the map, but subsequent reality may result in a modified classification based on what is feasible); (6) types of image data, photos, and other reference materials to be used; and (7) field methods to be employed.

Each of these components of objectives statements helps determine the methods selected for field work, including sampling procedures, locational techniques, and details of methods for making observations in the field.

Problem 2: Lack of a Valid Sampling Plan

Much of the planning for a field project must consider the difficulty of assuring the representativeness of field samples. Map accuracy depends greatly on the degree to which sampled data truly represent the land surface. This involves acquiring a sufficient number of samples in each category to be mapped and assuring that the aggregate of samples represents all the variation within each category. Failure to achieve this is one of the most frequent but preventable errors made in field work in remote sensing, and usually can be attributed to collecting too few samples.

Some methods of data classification used in remote sensing assume that data points have a random distribution over the study area. Often this assumption is ignored during collection of samples in the field, and the resulting map accuracy is compromised. If field data cannot be collected in a way that satisfies statistical assumptions of the classification program, then a less restrictive program should be used. The accuracy of the final map might be just as good with an alternative data classification program, but analysts should reveal that a program of less statistical rigor was used. Data analysts should understand how to choose a classification program and how to demonstrate that their data are suitable for that program.

Problem 3: Difficulty in Dealing with Scale Differences

This problem is one that initially overwhelms an inexperienced field person. The high resolution of the human eye at a distance of only a few feet presents such an abundance of ground information that one hardly knows how to relate it to the level of generalization on images and air photos. Field work consists largely in collecting information that can be scaled up by aggregation to correspond to information on images. To do this one must visualize the extent of a "ground pixel," which is the area of ground coverage represented by an image pixel. Then it is necessary to collect and aggregate ground data to best represent one or several image pixels.

Problem 4: Errors in Location

With a georeferenced image and a global positioning system (GPS) receiver, locational problems are greatly reduced, especially since the U.S. government's removal of selective availability, which occurred in May 2000. However, it is still difficult to be certain that a field location is accurately tied to a single specific image pixel coordinate. For this reason, it is necessary to estimate a potential locational error in pixel units and adjust the ground sample unit size accordingly. The potential for having a damaging locational error is highest on surfaces that have a high frequency of variation in cover type (e.g., urban areas). Large homogeneous areas allow some locational error without damaging effects as long as the location is not near a category boundary. Field of view (FOV) of the sensor also influences the precision with which location needs to be determined. Sensors with greater FOV allow more latitude for location errors than sensors having smaller viewing areas do.

Problem 5: Inappropriate Observations and Measurements

The question of what to measure, how to measure it, and to what level of detail it should be measured is still one of the greatest questions to field personnel. This issue is also given the least attention in many published articles. When searching theses and dissertations, one sees considerable attention given to measurement details, but when the work appears in a publication, measurement details are greatly abbreviated or missing altogether. The level of detail collected may be insufficient to meet the overall objectives in terms of the number of categories to be mapped or the level of accuracy targeted for each category. The opposite may sometimes occur when more data are collected than are needed to meet the objectives of the project. This mistake results from insufficient thought given to project objectives and can waste days of valuable field time. Some researchers prefer to err on the side of overcollection of data. They collect everything possible because of uncertainty about which biophysical variables are most significant in the reflectance of a surface material. A similar deficiency in data collection may result from measurement of features that have little or no influence on spectral response in the wavelengths sensed in making the image. For example, measuring water temperature when using visible and near infrared images would result in data that may have no relationship to the image. Any field project should begin with a brainstorming session to identify all biophysical variables that may affect spectral response in the wavelengths under consideration. The biophysical variables actually selected for measurement will be determined by reference to project objectives.

Some of the difficulty is a poor understanding of the relationship between biophysical variables and spectral responses of surface materials. The more a field person knows about the reflectance–absorption–transmission relationships of surface materials, the easier it is to select which biophysical variables to observe in the field. At the very least there should be an awareness of the basic responses of water, soil, vegetation, concrete, and asphalt to solar radiation in the reflective wavelengths. Further knowledge on the variations of these basic responses is valuable. For example, one should know the effect of turbidity on the response of water, the effect of moisture and texture on soil reflectance, and the effects of moisture, cover density, or biomass on vegetation reflectance.

Problem 6: Inadequate Reference Materials

Reference materials, other than field data, include all archival data such as air photos, maps, and any other compiled data that are referenced to map locations, for example, census data. The problem of inadequate reference materials may create more frustration and dilemmas than all other problems combined.

Reference data are considered inadequate when (1) scale and level of generalization of various maps and aerial photographs vary greatly and (2) dates of air photos, imagery, maps, and field work differ by time of year or by more than a few years. Project planners can overcome this difficulty by planning field work to coincide with overflights of satellites and aircraft, provided funds are available. Other projects must make do with poor synchronicity of reference materials by trying to minimize differences between dates or seasonality between reference materials and imagery. Acquisition of reference materials as well as a thorough search of the literature may actually reduce the amount of field work needed.

DEFINING OBJECTIVES

The importance of clearly defined and well-thought-out objectives cannot be overemphasized. A thorough definition of objectives requires considerable thought about each detail of a project and determines field procedures, level of generalization, sampling approach, data processing technique, and final product. In short, everything about the project should hang on the definition of objectives. Furthermore, the process is greatly helped by writing out the objectives. Plan on writing a detailed project objective statement as a necessary first step of project planning. The following components of such a statement should each be considered, although the sequence is not critical. An example follows the list of components, and although the example is a mapping project, the same elements would be considered in planning projects that generated something other than maps, such as a biophysical model or a validation of results of a previous project. The objective statements must always be thought out thoroughly.

Components of a Statement of Objectives

Tentative Title and Application of the Final Map

This may be the easiest and most obvious step in the preparation of a statement of objectives. The result here should be a name that is as specific as possible. For example, an agricultural survey might be called "Agricultural Land Use" rather than just "Land Use." If the survey was intended for irrigated crops only, then the title might be modified to "Irrigated Agriculture." Thinking about a specific title shapes, at a very early stage, the type of field work to be done. It helps one put an early focus on the features that need to be observed and which kinds of data will be gathered. How the user will apply the map may also be a part of the map title, for example, "Irrigated Agriculture for Estimating Water Consumption." This helps clarify a rationale for doing the work in the first place, as water managers would have an interest in applying water consumption of particular crops to a map of crop acreage.

Location and Size of the Study Area

An important determinant of location of a study site is often the availability of reference materials and image data. It is frustrating to select a study site and then find there are no data or adequate maps to use for reference. This problem may be unavoidable when project areas are selected by a client or other outside party.

An important factor in determining the appropriate size of a study area is the areal extent and uniformity of categories to be mapped. If a grassland or forest is being mapped, the typical low frequency variation may require a larger area in order to incorporate the necessary categories. On the other hand, an urban area presents a challenge at the opposite extreme in which even the smallest area contains a high frequency of variation both within and between classes.

There is a tendency among researchers to take on too much work for the available time and resources. Defining too large a study site is a common error. The optimal size of a study area is determined by the amount of time and money available, the number of people available to work in the field, the time required to collect data in the field, and the mode of travel possible in the field; for example, agricultural areas usually have many roads, but wilderness areas will require some travel by foot. Each of these variables should be thought out in detail and may have to be modified as the exact procedures for observations and data collection become more clear. Bigger study areas do not necessarily make a better project. The quality of project results may be improved by choosing a smaller area and working more intensively, rather than spreading effort over a large area with fewer data points. As with many issues in project planning, there is no single correct solution, only factors to be weighed.

One ever-present factor that must be considered is permission for access to a field site. If the proposed field site includes privately owned land, always ask the landowners or tenants for permission to enter. Be ready to explain in straightforward terms what the project is about and what a field crew will be doing on their land. Ask them which roads field personnel may use, and find out where to expect livestock. Assure them that all gates will be left as found, either closed or open. In some places strangers are not trusted, especially if they appear to be connected to government agencies. Advance contacts with local residents will ease their suspicions when they see strangers in unfamiliar vehicles in the area. If permission for access to crucial locations cannot be obtained, other study areas might need to be considered.

Probable Legend of the Final Map

As a practical matter, some project operators derive the map legend as a result of what is spectrally possible to map. Ultimately, reality overcomes idealism in the final stages of a project, and one must map what can be mapped. However, if the project begins with this approach, there is little to guide decisions on field procedures. Without a preconceived notion of a map legend, one runs the risk of measuring either too much or too little during time in the field. Remember that field time may be the most expensive element of the entire project, so use it wisely.

The map legend need not have the same level of detail in every category. For example, in a project mapping irrigated agricultural land use, dry land agriculture and settlements may each be a class without subclasses, while the irrigated land might have a category for each crop type, with subcategories indicating crop vigor and cover density. Identifying these differences in the map legend plays a useful role in planning what needs to be done in the field.

Obviously, some time needs to be spent in the dry farm area looking at the variation that must merge into a single class, but few measurements will be needed. In the irrigated areas one may need to observe crop type for each field, the stage of growth ranging from bare ground to ready for harvest, and the health of the crop. In addition, measurements may be needed to determine crop height and cover density. In this way the map legend becomes tied to field procedures. If one finds that certain intended classes are so similar spectrally that they cannot be mapped by the available remote sensing method, then class merging will be necessary and the legend will be altered. This decision may be made at a very early stage when first viewing the result of a cluster map of the image data for the study area.

Map Scale and Level of Map Accuracy

It may sound too presumptuous to attempt to set a target for map accuracy before the project even begins. However, this issue is not a matter of wishful thinking and ambitious goals for a perfectly accurate final map. Rather, it is a matter of thinking about the relationships among map scale, level of generalization, and accuracy. As a general rule of thumb, as map scale becomes smaller (larger areas), mapping units (cells) and categories are aggregated, causing map generalization and accuracy to increase. Keep in mind that one 20-meter (m) pixel of Thematic Mapper (TM) data will appear as a spot 0.83 millimeters (mm) wide on a 1:24,000 scale map. Imagine the headaches of field work if one attempted to produce a final map with this level of detail. The resulting map accuracy would likely be very low, assuming it is possible to find pixel-sized locations in the field precisely enough to match with image pixels to assess accuracy.

As mentioned above, the number of map categories and the homogeneity of cover within categories also affect map accuracy. Unfortunately, there is no good rule of thumb for predicting final map accuracy, but there are some elements to consider. If there are six or fewer map categories in the legend, and each is somewhat homogeneous, overall map accuracy of 90% or better is a reasonable expectation with comparably high accuracy in each category. As the number of map categories and cover variability increase, map accuracy, overall and by category, will decrease. In a complex surface with many map categories and a map scale of 1:24,000 or larger, plan for an overall map accuracy of 65–70%. The final result may be better. As we will see later, the number of field samples needed for classification depends on the expected level of accuracy.

Types of Image and Reference Data to Be Used

Ideally the selection of image data is based on the objectives described up to this point. The appropriate wavelength bands, and the resolution (spatial, radiometric, spectral, or temporal) for detecting and mapping the phenomena in question determine the image data selection. Seasonality also often influences the choice of image data.

The selection of a study area is often influenced by the availability of reference materials, including air photos, existing topographic maps, cover type maps, soils maps, census data, and maps. Acquire everything.

Preprocessing and Classification Approach

This guidebook is not intended to discuss classification techniques, but since the issue should be mentioned in a statement of objectives, a few comments are in order. First, know the statistical structure of the image data, and select classification algorithms whose assumptions are not seriously violated by the data. Keep in mind that the maximum likelihood classification method, though quite rigorous, assumes a normal distribution of data. If necessary, consider data transformations as part of the preprocessing to make the data better match the assumptions of the classification algorithm. Second, consider a layered classification approach, beginning with cluster analysis followed by a supervised algorithm. These issues are thoroughly discussed in Jensen (1996). Many remote sensing personnel use the classifier that gives the best result regardless of the data structure. If the results are optimal, perhaps the statisticians' advice can be ignored with a clear conscience.

A Statement of Objectives

The following example states the objectives of a study of irrigated agriculture land use, as mentioned above. This statement is for use in project planning. An actual project proposal would, of course, elaborate extensively on methods and other details.

Example

In order to learn more about the consumption of water for irrigation, it is important for water managers to have reliable information on the types of crops and their acreage (shows application). The objective of the project is to map the crop type, stage of growth, cover density, and acreage of irrigated agriculture in White County within the limits of the South Fork 1:24,000 U.S. Geological Survey (USGS) topographic map (shows location, map coverage, and scale). The final map will display categories including sugar beets; corn, new growth; corn, mature; alfalfa, recent cutting; alfalfa, mature; bare ground; pasture; dry farmland; settlement (shows legend). Based on a minimum unit area of 10 acres at a scale of 1:24,000, and a random sample of measurements taken within each field, the final map will strive for an 85% overall accuracy and 80% or better accuracy for each category. Each 10-acre unit will show the dominant cover type for that location (shows expected accuracy and cell size). Accuracy analysis will be done by field work and air photo analysis to produce an accuracy matrix. The primary data will be Landsat TM selected to cover the area in late August before harvest near the end of the water consumption season. Reference data will consist of field updated air photos of a date as close as possible to the TM data. Other reference data will consist of crop data from the White County agricultural agent, topographic maps, soil maps, and water allocations for each parcel of land (shows image type, date, and reference sources). Data processing will consist initially of atmospheric corrections in each band of the TM data set. Classification of TM data will begin with cluster analysis, followed by comparison with reference data for class merging. Final classification will be done by a minimum distance to means algorithm (preprocessing and classification approach).

PLANNING FIELD WORK BASED ON OBJECTIVES STATEMENT

Assume that it is now spring or early summer and time to select TM overflight dates for late August or September prior to harvest time. Field work can be divided into tasks to be accomplished before the TM overflight, during or near the time of TM overflight, and possibly work to be done after the TM overflight.

Field Work before Overflight

Before going to the field, remote sensing personnel should investigate the availability of aerial photographs and satellite imagery, and place orders as needed. TM data may be ordered from Earth Observation Satellite Company (EOSAT) at 4300 Forbes Blvd., Lanham, MD 20706; 800-344-9933. Also practicing field measurement methods with instruments is important. This preliminary work trains field personnel in the methods and use of instruments, and determines whether all equipment is operating properly. Contingency plans should be made to cover unexpected events, such as bad weather or equipment failure.

Considerable effort may be required to update air photos. If the project has resources sufficient to pay for air photos taken concurrent with a TM overflight, then an air photo update is not necessary. If the air photos are from some previous year, then this task requires going to the field after crops have begun to grow in order to identify the changes in fields that have occurred from the time of the photo overflight. For example, some fields that are bare or fallow in the air photo may have a crop in the current year.

Field Work during or Near the Time of Overflight

Identify crops and make appropriate physical measurements for each category of irrigated crop at randomly identified measurement sites. Identify land cover in dry farm areas at sample sites for each crop type. Identify settlements and other built developments.

Field Work Done after Overflight

In this example, all measurements must be made near overflight time because it is scheduled just before harvest. After harvest any field work that involves measurements will be of little use. Details of permanent developments or settlements may be observed or measured anytime before or after an overflight.

This example demonstrates the use of preparing a complete statement of objectives as an aid in planning field work. The rest of this guidebook is designed to help in the execution of the field work and will provide details for some of the points covered in the statement of objectives.

2

Sampling
in the Field

Following an appropriate sampling strategy is as important as the actual collection of data from the field. This chapter presents the elements to be considered in selecting a sampling strategy.

It is imperative to plan the sampling strategy carefully, as it, more than anything else, will determine the amount of time spent in the field, the accuracy of the results, and the confidence in the final map. In order to be effective, field samples must be representative of all the variation contained within each information class. Five basic decisions must be made in devising a sampling strategy: (1) selection of aerial photographs and mapped reference materials; (2) timing of the sampling process relative to performing the classification; (3) sample site configuration or pattern; (4) number of observations (sample sites) to make; and (5) size and spacing of sample sites. These issues apply to samples taken for both training and accuracy assessment.

SAMPLING FROM AERIAL PHOTOGRAPHS OR THEMATIC MAPS

Aerial photographs should be used for sample site selection only if they are close to the image in date, or if it can be determined that cover materials have not changed over time. Even if the photos are close in time to the images, field visits should be made to ascertain the accuracy of photo interpretation of the cover types. If both the image and the photo are from an earlier time, the current cover types in the field may be different from both, especially in agricultural areas or urban fringes where change is frequent. In such cases, field work will be of limited value and the project can be done in the lab. If only cover type is needed with no field measurements, aerial photographs may be completely adequate.

Maps of cover type are a poor substitute for photographs and should not be used for training or accuracy site selection unless there is no alternative. Maps are by nature generalized information and likely will have used different definitions of categories, as well as a different minimum cell size, than is being used in the current project. Either of these deficiencies makes a map unreliable and invalid for use as a reference material, other than for general knowledge of the area. Many maps do not provide sufficient information on category definition and minimum cell size to determine whether they might be compatible with a different project. It is best to avoid the use of thematic maps for training or accuracy site selection.

On most present-day image analysis software it is possible to select training data on the computer screen by creating training polygons, or by a seeding process in each area that appears likely to be a class. If this approach is taken, it is important to go to the field to make on-site identification of the correct labels for the classes or to make measurements of biophysical properties for sites selected on the screen.

TIMING OF OBSERVATIONS

An important consideration is the time of field work relative to the overflight of sensors collecting the image data. The ideal is to have field sampling coincident with the overflight. The more dynamic the features being mapped (e.g., vegetation in the growing season), the closer in time the field work should be to the time of the overflight. Cloudy weather often makes this ideal plan difficult. If long-range plans must be made, it may be best to plan field work a day after the overflight. This way, field personnel will know before beginning field work if the area was overflown, if an image was obtained, and whether the image is cloud-free. If the overflight day is cloudy in the field area, then field work can be rescheduled to correspond with a different overflight. Even a 1-day delay in field work may create problems if sampling highly dynamic phenomena such as soil moisture, tillage or harvesting operations, or changing water levels in lakes and rivers.

Another question of sampling timing is concerned with whether to collect field data before or after the classification stage has been completed. The answer to this timing question depends on whether the project will use a supervised classification, an unsupervised classification, or a combination using unsupervised clusters as training sets for a supervised classification.

The supervised classification approach requires field work to identify training sites in accordance with a predetermined list of categories, or potential map legend, before image processing. In this case, it is necessary to create a georeferenced image prior to field work so that coordinates can be provided for selected field sites. The issue of field site location will be discussed later.

The unsupervised approach utilizes field work or photo analysis after the initial clustering process to identify the map category (information class) represented by each cluster (spectral class). If biophysical measurements are needed, then that work may be required for each cluster. This approach ensures that all the spectral variation within the information class is represented. Observations in the field will determine whether clusters should be merged or stand alone as information classes. Figure 2.1 depicts the relationship between spectral classes and information classes. In Figure 2.1 note that several clusters may be part of one information class and that more than one information class may be represented by only one spectral class. Occasionally clusters and information classes may coincide.

SAMPLING PATTERNS

Five basic patterns may be considered for sampling in the field: (1) simple random, (2) stratified random, (3) systematic, (4) systematic unaligned, and (5) clustered. Another approach, purposive sampling, though often used, lacks any structure or systematic plan, so it cannot really be called a pattern. These sampling plans are not equally valid for all remote sensing projects. Terrain factors may play a role in selection of an appropriate sample plan. Numerous

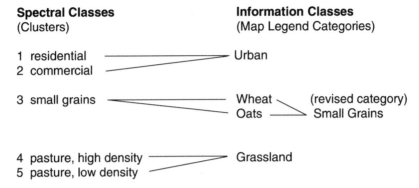

FIGURE 2.1. The relationship between spectral classes (clusters) and information classes in a hypothetical project. Identification of spectral classes in the field determined that clusters 1 and 2 will merge into a single class, and the same is true for clusters 4 and 5. Because of the similarity in spectral responses for wheat and oats, only one cluster exists for two categories in the map legend. This would necessitate revising the map legend to show the more general term "small grains" rather than trying to include both wheat and oats.

books on spatial statistics for earth scientists provide a discussion of sampling patterns. Useful resources are Williams (1984), Silk (1979), and Justice and Townshend (1981).

Simple Random Pattern

The purpose of a simple random sampling pattern (Figure 2.2) is to ensure that all parts of the project area have an equal chance of being sampled with no operator bias. This condition is important to the assumptions of the underlying statistics used in classification. Random sites are selected by dividing the project area into a grid with numbered coordinates. Then coordinate pairs are selected from a random number table and plotted on the project map. Each random point becomes a sample point or the center of a sample area. The size of grid cells is an important consideration that applies to any of these sample patterns and will be discussed later.

Although the simple random pattern leads to a minimum of operator bias, there is a serious drawback to it as a sample pattern. A random pattern over the entire study area is not likely to have a uniform distribution of points, and categories having small areas may be undersampled or missed entirely. Also, some points may prove to be inaccessible.

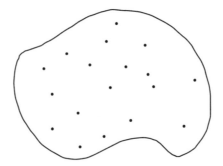

FIGURE 2.2. Simple random sampling pattern.

Stratified Random Pattern

A stratified random sample pattern (Figure 2.3) maintains a necessary randomness and overcomes the chance for an uneven distribution of points among the map categories. This approach assigns a specific number of sample points to each category in proportion to the size or significance of the category with regard to the project objectives. If all categories are of equal significance to the project, then category size alone determines the number of samples in each. To maintain a random pattern, points should be assigned within categories using a grid and random numbers, as in the simple random pattern described previously.

Systematic Pattern

A systematic sampling pattern (Figure 2.4) assigns a sample point to positions at equidistant intervals, for example, all grid intersections. Although the ori-

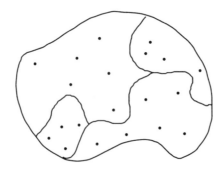

FIGURE 2.3. Stratified random sampling pattern.

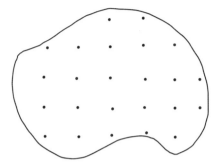

FIGURE 2.4. Systematic sampling pattern.

entation of the grid may be chosen randomly, because every position is determined by the choice of starting point, randomness is not achieved. This method does not satisfy the assumptions of randomness when using inferential statistics. Another drawback to this method is that such a regular orientation of samples is likely to introduce bias due to some linearity of patterns in the landscape, particularly in regions with a rectilinear land survey system.

Systematic Unaligned Pattern

The systematic unaligned sampling pattern (Figure 2.5) uses a grid, as in the systematic method described above, but assigns the position of each point randomly within the grid cell. In this way a degree of randomness is maintained within the constraints of the grid cell, but the grid assures that all parts of the project area will be sampled. In order to avoid the tedious operation of devising a pair of random numbers for each cell, it is acceptable to determine

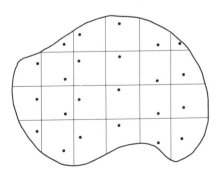

FIGURE 2.5. Systematic unaligned sampling pattern.

a random number for each column and for each row of the grid. This approach will result in a random location in each cell.

Using the systematic unaligned sampling pattern assures that sample points will be evenly distributed over the study area and that all classes will be represented. The random position in each may be sufficient to overcome unintended alignments of sample points and landscape features. Whether this is true depends on the size of the cells in the grid. If cells are small relative to landscape alignments, there may still be a problem as described in the systematic sampling pattern. Hence, the crucial issue here is finding the appropriate cell dimension for the grid. Figure 2.5 shows that, by decreasing the size of the cells, a systematic aligned pattern would eventually occur. For example, if there needs to be a large number of samples in order to improve the expected accuracy or the level of confidence, then it would be necessary to have a larger number of smaller cells in the grid. In this situation the element of randomness could be essentially lost.

Clustered Pattern

In the clustered sampling approach (Figure 2.6), nodal points are used as centers for clusters of sample points radiating from the center. Any number of nodes can be selected, and any number of satellite points can be tied to them. Nodal locations can be selected randomly, stratified by category or selected strictly by identification of accessible sites. Furthermore, the satellite points can have random directions and distances from the node even if the node itself was not selected randomly.

From a practical point of view, the clustered sampling approach has some important advantages to a field survey. In terrain with poor access, this

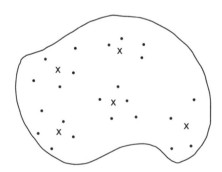

FIGURE 2.6. Clustered sampling pattern.

method allows the operator to make the most of accessible sites. Another advantage is that field time may be greatly reduced by having fewer sites (nodes), hence less travel time, and multiple satellite sample sites may be visited while at each nodal location. By imposing randomness on the selection of nodes and satellite sites, the assumptions of inferential statistics can still be met. However, the satellite sample sites must be far enough from the nodes to overcome the problem of autocorrelation, thus giving a false statement of map accuracy. Since there is no rule of thumb to follow to determine the distance that a satellite sample site should be from its node, one must be sure that satellite sites are sufficiently far apart to overcome autocorrelation and to make sure that variation within the category is represented. The possibility of autocorrelation is high enough to recommend avoiding the clustered pattern unless other methods are too impractical because of terrain problems.

Purposive or Judgmental Sampling

Purposive sampling is based entirely on the operator's judgment in purposely, or deliberately, selecting "representative" sample sites. This is defensible if the operator is thoroughly experienced in working with the phenomenon being sampled and is also very familiar with the extent of variation in the study area. The field person's experience determines where sample sites should be in order to best represent the variation within a category.

The advantage of this approach is that representative sites might just as well be reasonably near roads rather than trekking to a distant site to get the same type of information. However, only a person very familiar with the area would know that the distant site indeed provides the same type of information. This method may be the best option in areas where accessibility to some parts of the study area is a serious problem.

A serious disadvantage is that operator bias is always present, no matter how carefully the work is done. Therefore, statements of accuracy and level of confidence in the final map are statistically invalid, even if the numbers are high, because randomness was not employed. When selecting sample sites purposively, the credibility of the accuracy statement depends entirely on the reliability of the person making the observations. Nevertheless, because of the practicality and ease of site location, this method probably has fairly high usage despite its poor recommendations. If the end user of the project map is also the person most familiar with the area, it may be well to have that person involved in the field when using purposive sampling.

Anyone opting to use the purposive sampling approach should consider taking enough photographs of sample sites to demonstrate that the range of variability in each category is represented. When using purposive sampling, care should be taken to use classification programs, such as minimum distance to means, or spectral angles (Sohn & Rebello, 2002), which do not assume randomness of data points.

COMMENTS ON SAMPLING PRACTICALITIES

Some practical considerations must be introduced to the selection of sample sites. One issue is the selection of an appropriate sampling procedure. Another issue is the acceptance and rejection of certain randomly selected sample sites.

There are two questions regarding rejection of randomly selected sites:

1. What rules should be followed for rejecting samples?
2. If undesirable points are thrown out, is the pattern still random?

Even at the risk of compromising randomness slightly, it is absolutely necessary to reject certain randomly selected points. The most important rule is to reject points lying on boundaries between categories. Every sample site must be wholly within a category. Other criteria for point rejection are more subjective and should be used with caution. For example, as a practical matter, it may be necessary to reject some points because of inaccessibility. However, this rationale can easily be misused and lead to rejection of points that are merely inconvenient. Overuse of this practice would eventually lead to a nonrandom group of sample points. One can imagine carrying this rejection process to an extreme, resulting in a set of "random" samples all located within 200 feet of roads. Any rejected sample point should be replaced by another randomly selected point.

It is important to be clear about sampling procedures when writing the project report. Professional journal articles often omit these details, causing much frustration for project planners looking to the literature for guidance. Since there are a variety of approaches, each with its advantages and weakness, operator integrity requires a frank description of any factors that may contribute to weaknesses in the results. At the same time, the researcher should explain reasons why the methods chosen are well suited for the situation. This openness shows an awareness of the advantages and disadvantages associated with the selected methods and provides a justification based on trade-offs between propriety and practicality, assuming that practicality has not dominated. One must realize, also, that no single correct sampling approach applies to all situations. Tests by various researchers have shown that simple random and stratified random patterns both give satisfactory results. However, the stratified random approach requires some advance knowledge of where the strata boundaries are. This is not always possible, but often the probable boundaries can be ascertained with aerial photographs and visits to the field. If unsupervised classification is used as a preliminary step toward supervised classification, the class boundaries are known at the beginning of sampling. This makes it possible to establish a stratified random sample pattern for both the training and the accuracy assessment stages.

NUMBER OF SAMPLES

Training Sites

The most important guide for selecting sample sites for training data is to be sure that all the variability within classes is accounted for. In some homogeneous classes this could require only a few sites. More sites will be needed in classes with high variability. A general rule of thumb offered by Jensen (1996) is to select a number of pixels in each class that is at least 10 times the number of bands used in developing training statistics. For example, if 6 bands of TM data are used, then at least 60 training pixels are selected for each class. This is sufficient to allow valid computations of variance–covariance matrices usually done by classification software. In the discussion below on size of sample sites, it is noted that as sites grow larger than 10 pixels, there may be no new information added. Therefore, it would be better to have six sites of 10 pixels in each class rather than one training site of 60 pixels in each class.

Accuracy Assessment Sites

In spite of efforts by various researchers, there is still no hard and fast rule for determining the number of samples needed for accuracy assessment. However, there are some good guidelines. One suggestion (Fitzpatrick-Lins, 1981) applies the following binomial probability formula to estimate an appropriate number of samples over the entire study area:

$$N = Z^2 \, (p)(q)/E^2$$

where N is number of samples, $Z = 2$ (the standard normal deviate for a 95% confidence level), p is expected accuracy, q is $100 - p$, and E is allowable error. For example, if an accuracy of 85% is expected, with an allowable error of 5%

(95% confidence level), 204 sample sites would be needed to apply this approach. If the allowable error changed to only 2%, the number of samples needed would jump to 1,275. However, this method does not consider the size of the study area, the number of categories, the variation in the areal size of the categories, nor the variability within the categories. It would be very easy to undersample using this method alone unless the confidence level is set to a point that is unrealistic in terms of field time.

A suggestion by Congalton and Green (1999), based on experience with the multinomial distribution, is to use 50 samples minimum in each category. If the area is larger than 1,000,000 acres, or if there are more than 12 categories, then there should be 75–100 samples per category. This approach samples small areas heavily, while large areas might be undersampled. These suggested sample numbers could be adjusted to accommodate variations in size and within-class variability.

Evaluation of accuracy is an important, though often omitted, part of any project. Although data collection for accuracy assessment may be a job for field personnel, the evaluation is done in the laboratory by means of an error matrix, sometimes called a confusion matrix. This procedure leads to an understanding of accuracy of classification for each category, as well as overall accuracy. Excellent sources for developing a valid evaluation of accuracy include Campbell (1996), Jensen (1996), and Congalton and Green (1999).

SIZE OF SAMPLE SITES

Although the term "sample point" is often used, seldom is it possible to collect data at a single point to represent an area. Point data can be used only when a high degree of generalization is acceptable or when there is no variation in the landscape over an area several pixels in size. For example, point data on a water body is appropriate in projects that are not concerned with variations within a category like water. Be guided by the spectral classes produced by a cluster analysis to assess the uniformity of an information class. Be aware, however, that a given pixel may change its cluster membership if subsequent cluster analyses have a change in some input parameter, such as maximum number of clusters. Several runs of the cluster analysis program will give an idea of the relative stability of each category. Pixels that consistently occur together in the same cluster have good stability and indicate a strongly homogeneous category.

The main things to keep in mind in determining the size of a sample site is the variability of the terrain and the ground resolution of the image data, which is usually the same as image pixel dimensions. Remember that the objective is to find a relationship between the single data values in each band of an image pixel and the great amount of information and data in the corresponding ground pixel. Except for applications involving pixel unmixing (subpixel analysis), one will try to generalize a great amount of variation on the ground into a single statement (either qualitative or quantitative) that is representative of an area at least the size of a ground pixel, though usually larger.

In order to associate a particular ground pixel with a specific image pixel, the field person must have precise knowledge of the geographic coordinates of the sample site. Fortunately, satellite global positioning systems have made this task much less difficult than in the past. With selective availability deactivated, location accuracy of approximately 15 meters can be obtained routinely with a hand-held GPS receiver, even without differential correction. Depending on the configuration of satellites at a given time, accuracy may be even better.

With GPS should a field person try to have sample sites of only one ground pixel in area? This minimal area is not advisable because of errors on the order of 0.5–1.0 pixels, which usually exist in georeferenced images, in addition to errors in ground location. Justice and Townshend (1981) suggest a useful formula for determining area of a sample site:

$$A = P(1 + 2L)$$

where A is minimum sample site dimension, P is image pixel dimension, and L is estimated locational accuracy in number of pixels. Suppose, for example, that a project is using TM image data with 30-m pixel dimensions, and one can accurately locate an area on the ground to ±0.5 pixel (15 m). The minimum ground area dimension that should be sampled is 60 m × 60 m. Sample sites may be larger than the minimum to allow for a margin of error in ground location and image georeferencing. If operating a GPS without differential correction and without selective availability, a field operator should not estimate a locational accuracy (L) better than 15 m. Keep in mind that the coordinates of a georeferenced image might also have a 15-m error (0.5 pixel) in TM images. One cannot know the direction of displacement of both the GPS and the image error. They might be additive and create greater total error. Only in areas with a high spatial frequency of variation in surface cover is there a need to strive for minimum sample dimensions. This leads to the consideration of surface homogeneity as a variable in sample site dimensions.

Homogeneity does not mean there is no variation within a sample, but rather that the variation is uniformly distributed over the area and will produce a unimodal distribution of data. Evaluation of uniformity is very subjective, but Joyce (1978) gave some guidance that is portrayed in Figure 2.7. A more practical and workable guide to homogeneity is the spectral classes produced by cluster analysis. One can assume that each spectral class (cluster) is more or less uniform and is different from other spectral classes. Keep in mind that the cluster maps can vary within the same data set, depending on the parameters set for the clustering process.

Where does this leave us after considering all the variables affecting sample size? How can one best choose a size for sample sites? The practical approach is to identify sample sites based on the size and shape of spectral classes of a cluster map. Sample areas of variable size can be identified within each spectral cluster and located in the field. It is easy to see if a cluster is large enough to locate a sample site safely without danger of location error. As a minimum, sample units should be no smaller than a 3×3 cluster of pixels or a polygon of comparable size for either training sites or accuracy assessment sites. On the other hand, if sample sites are larger than 10–15 pixels, little new information is being added and field time may be wasted. Sample sites containing more pixels add to the amount of field work required at each site. More information is added by having numerous sample sites of up to 10 pixels each in size.

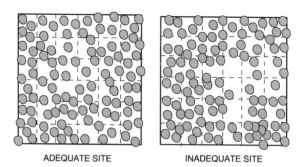

ADEQUATE SITE   INADEQUATE SITE

FIGURE 2.7. The uneven distribution of trees in the figure on the right makes it a poor sample site. The same applies whether considering plants of the same species or an uneven distribution of species in a mixed community. Adapted from Joyce (1978).

WHERE TO BEGIN MEASUREMENTS

In order to preserve randomness during field work the field person should choose a method for removing bias from the selection of both a starting point and the direction to so in measuring features at specific points along a line, or transect. For example, one might assume that the starting point is the center of a randomly selected sample site. If measurements are to be made along a transect from that point, one should consider a method for selecting a direction for the transect. This might be done by randomly choosing a compass bearing, or by simply flipping a pencil and letting it point in the direction of the transect. This helps avoid choosing a line that is easy to traverse but which may not provide representative data along the way. It would also be valid to select the direction north, or any other direction, and use it for starting a transect at each sample site. This would be sufficient to prevent the field operator from introducing bias in selecting a direction to begin work. The field person may need to select several transects within each sample area in this manner. Note the transect pattern suggested by Joyce (1978) in Figure 2.8. This approach is valid, and some such approach should be followed consistently at each sample area.

If areas, or quadrats, are being used as the basic site for collecting field data, field personnel should have a plan for locating quadrats to be spread over the sample area. For example, one might choose to measure a fixed number of quadrats spaced according to a predetermined plan over each sample area. The important element of any field work, if randomness is required, is that no bias should be involved in selecting features to measure.

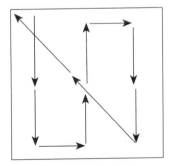

FIGURE 2.8. This suggested transect pattern covers all parts of the sample site and brings the observer back to the point of beginning. The length of the transect will vary according to the size of a given sample site. Adapted from Joyce (1978).

3

Finding Locations
in the Field

Any remote sensing project that requires field measurements, or site-specific observations, must have a means of determining reliable locations for each site. Location in the field before the introduction of a GPS sometimes used surveying techniques such as creating a triangulation net connecting all sample points, but most often relied on identifying landmarks on the ground that could also be seen on the image (e.g., bridges, intersections of roads or streams). The field person had to take samples adjacent to, or a measured distance from, the landmark. This restriction greatly hampered taking samples in remote areas with few landmarks unless the area to be sampled was large and fairly homogeneous, allowing for large errors in location. One simply accepted the possibility that locational accuracy would be weak.

In urban areas, or well-developed agricultural areas, the locational problem was less restrictive due to the abundance of easily identified landmarks. Using those early techniques for location in the field often required that a large number of contiguous pixels be used for each site, whether used for training, field sampling, or verification. Field personnel still find it useful to know how to follow a bearing with a compass and to make measurements by pacing, taping, or using devices such as the hip chain, which uses an unwinding string to record the distance a person walks. However, today's GPS technology is absolutely vital to finding locations in the field and must not be ignored. Every researcher who plans to do field work in remote sensing must have access to a GPS receiver and learn to use it proficiently.

The advent of GPS technology in the 1980s and the completion of the satellite array in 1995 have greatly changed the approach to location-finding in the field. With selective availability deactivated since May 2000, location accuracy of 15 m or better is obtained routinely 95% of the time with a handheld GPS receiver and without using differential correction. Depending on the geometric configuration of satellites at a given time, estimated accuracy is often much better than 15 m. An inexpensive GPS receiver can enable remote sensing field workers to navigate with confidence to a point on the ground with known coordinates derived from a map or georeferenced image. Or it can help them determine such coordinates for sample points selected in the field. Both of these functions have important applications in remote sensing projects and will be discussed in detail later in this chapter. What will not be covered in depth is the actual operations of the GPS. This is because the purpose of the chapter is to acquaint field personnel with field procedures using GPS. For greater detail on system operation, recommended readings are Geomatics Canada (1995), French (1999), VanSickle (2001), and Hurn (1989).

OVERVIEW OF LOCATION-FINDING USING GPS

The U.S. Department of Defense initiated the system, called NAVSTAR GPS, which provides the best means available for navigation and position location. The system is an array of 24 orbiting satellites that emit radio signals, including three spares for backup. All the satellites are in 12-hour orbits at just above 20,200 kilometers (km; 12,500 statute miles) above the surface of the earth, and are arrayed so that at least four are available for use at any location between about 80° north and south latitude. Four satellites are the minimum needed to obtain reliable positions and elevations. It is desirable to have a receiver that will track at least eight satellites (some will track 12) so that the best four can automatically be selected for position computation. If one of the four satellites deemed optimal for location computation should move out of view over the horizon, another is already on track to fill in as necessary.

When a GPS receiver is turned on, it first downloads a current almanac of all the satellites, whether in view or not. The satellites that are in view may then be shown on the GPS receiver screen. Usually the display is on an azimuthal grid with the center at the user's location. Also, there may be a display showing signal strength either as a bar graph or as a number next to the satellite plotted position. Each satellite has a unique signal code that identifies it to the receiver on the ground. If a receiver has not been used for several months, or is more than 300 miles from its last position, it will need to initialize and may request operator input on the approximate present location.

One may obtain updated satellite almanac information on the positions of satellites, including satellite visibility and geometry before going into the field. Almanacs and other information on the satellites in the NAVSTAR constellation can be obtained on the U.S. Naval Observatory website, *http://tycho.usno.navy.mil/gps.html*. Select "GPS Interactive Satellite Visibility," and input latitude and longitude of the field area of interest and the expected date of the field trip. The result will be a list of optimal satellites at that time and place. One can then try some alternate dates and hours to compare with the first results. With this almanac information, the user can create a chart for a given location showing the days and hours of optimal satellite coverage. Figure 3.1 is an example of such a chart. A blank azimuthal chart is provided in Appendix 2.

Two means of taking position observations may be utilized by field workers in remote sensing single-point location and differential location. Differential location requires that data be collected from two locations simultaneously. Although differential location can achieve submeter accuracy, its use is limited in remote sensing projects if navigation to a known point is needed. However, when points are selected and their positions recorded in the field, differential location correction may be warranted if accuracy needs require it. For most purposes, single-point location will provide adequate accuracy unless the signal degradation called Selective Availability is turned on again sometime in the future.

Single-Point Location

Location of a point operates on the principle of intersecting circles marking a point (Figure 3.2). The circles represent all equidistant points between each satellite and the GPS receiver on the ground. The intersection of the circles determines the actual location of the receiver. The diagram in Figure 3.2 is a two-dimensional representation that does not show the effects of tiny, but significant, clock error in the receiver instrument. In practice, a fourth satellite

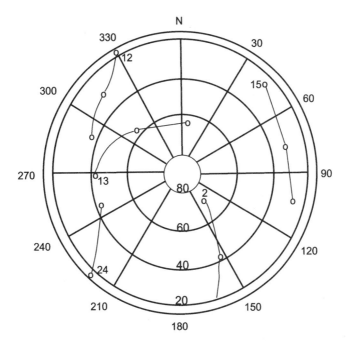

FIGURE 3.1. Radial lines represent azimuths and concentric rings represent elevation angles. The outer ring represents a mask that screens out any satellites less than 15° above the horizon. The paths of five satellites, identified by number, are plotted over a 2-hour period. Not all GPS equipment has the capability for making these plots, but with information on the U.S. Naval Observatory website such plots can be made by hand. Adapted from Geomatics Canada (1995). Used with permission.

signal is needed to compensate for small time discrepancies in the receiver instrument's clock. Hence, a minimum of four satellites is required to determine a location.

Differential Location Measurements

Greater location accuracy requires a stationary GPS station that records data continuously while a roving receiver takes data at selected points in the field. Fixed stations are operating continuously in many places now, and may be located as much as 300 miles from the roving receiver. Check with a local surveyor's office for information on operational base stations in the area of interest. When the field personnel return to home base they can download into a

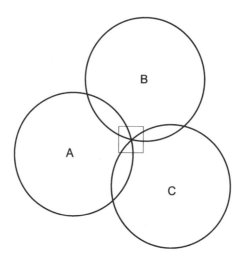

FIGURE 3.2. This two-dimensional version of the geometry on which position computations are made shows that signals from three satellites define a point location. The distance from satellite *A* applies to any point on circle *A*. The same applies to circles *B* and *C*. The intersection of the three circles, shown within the square, is the only correct location. The NAVSTAR system operates in three dimensions, and location is computed on the intersections of spheres rather than circles. The intersection of three spheres forms two widely separated points, but one of them is easily identified as an incorrect solution. Adapted from Hurn (1989). Used with permission.

computer both the roving data and the fixed station data for the time period they were in the field. Software obtained with certain receivers will make the appropriate signal matches and compute differential corrections for all locations obtained by the roving receiver. An explanation of the details of this procedure would be specific to various manufacturers' software and is beyond the scope of this book.

Sources of System Error

Inherent in the NAVSTAR GPS system are errors that include orbital errors, satellite clock errors, ionospheric path errors, receiver clock errors, and receiver noise related to instrument design. These sources of error are beyond the control of the field person during a field survey, but the operator can control for other possible error sources.

Errors That Can Be Avoided

The operator is able to avoid or mitigate multipath errors, interference errors, poor satellite geometry, and obstructions to radio signals.

Multipath errors occur when a radio signal from a satellite travels directly to the receiver but also indirectly by reflecting off a nearby feature such as a building, cliff, or other high landform. Hence, the operator is unaware that two or more conflicting signals are received for the same location. The magnitude of multipath errors will increase as the path length of the reflected signal increases. If the operator is not able to move away from the reflecting feature, there should be a note made of any potential for multipaths related to nearby buildings or landforms.

Interference errors—those that adversely affect radio transmission—may occur when the GPS receiver is near electrical fields, such as power lines, or in electrical storms. Whenever possible the operator should avoid collecting data near power transmission lines or during thunderstorms. If there is no choice, the operator should make note of possible interference and recognize a potential for error at that location. Other sources of interference include microwave transmitters and radio repeaters.

It is easy to see that the geometry of the satellite positions is critical to the accuracy of location. Imagine, for example, the result of using triangulation when the angles involved are either very small or very large. This would be the case if all satellites were nearly directly overhead or at opposite horizons at the time of taking measurements. Most GPS receivers compute an estimate of position error resulting from satellite geometry alone. The computed error estimate is usually shown on the receiver display screen as a distance value in feet or meters. Remember that this expression of error is based only on satellite geometry. If satellite geometry is inadequate, the receiver may simply refuse to compute locations. Fortunately, this problem is seldom experienced, but if it occurs there is not much the operator can do but wait for a change in the geometry of the satellite constellation.

Optimal satellite geometry for both accuracy and signal reception is to have one satellite nearly overhead and at least three others distributed evenly over the sky. If the receiver is capable of tracking more than four satellites, it will repeatedly compute an error estimate for each combination of four available satellites and use the four that give the lowest error term for each location computation. This is the main advantage of using a receiver that can track at least eight satellites. If the receiver can find only three satellites for computation, it will switch from the three-dimensional (3D) mode to the two-dimensional (2D) mode, and accuracy will degrade considerably, with results that may not be usable.

Obstructions to line of sight between a satellite and the GPS receiver will remove that satellite from the constellation visible to the receiver. If there is good satellite availability at that particular moment, the obstruction may make no difference, but the operator should make note of the obstruction and the potential for an erroneous location measurement. Obstructions to look for include buildings, landforms, and even the operator's own body if the antenna is held closely while facing away from useful satellites in the constellation.

Tree canopies may act as a partial obstruction that may attenuate the signal from a satellite. In a dense forest it is possible to lose contact with satellites altogether, but as antenna design improves, forest canopy is becoming less of a problem for accurate location measurements. The operator should be aware of potential errors under trees and make note of such. Also, the operator should keep an eye on the estimated accuracy information provided by the receiver during location measurements. If error estimates become significantly larger under tree canopies, it means that some satellites have been lost to view, and adjustments should be made in the size of the field sample site to accommodate uncertainty in site location. The estimated accuracy depends on the geometry of the satellite array at the moment and the strength of the signal being received. Therefore, if anything is attenuating the signal, the estimated accuracy will degrade.

When the remote sensing field worker sees a potential for significant error in site location measurement, it may be useful to average measurements over a period of time. Even though averaging may slow the progress of the field work, its advantages far outweigh the loss of time if errors are lessened in the process. Many receivers are now designed to average readings on request. Normally this automatic averaging takes only a few seconds longer and may provide better location accuracy. Having measurements taken every 2 minutes over a 30-minute period and averaged might be worth it if accuracy at a particular location is vital to the project. This practice would be warranted only if Selective Availability were turned on.

When moving from site to site by automobile, an effort should be made to keep the receiver antenna in view of the sky. Usually mounting the receiver or a remote antenna on the dashboard next to the windshield can do this. Keep in mind, though, that in this position the roof of the automobile may obstruct some of the available satellites, forcing the receiver to select satellites that may be less than optimal for that time. After transporting the receiver inside an automobile, the operator should allow a short time for the GPS receiver to reorient to all available satellites upon arriving at the destination site before determining a location. If the receiver instrument can accommodate a rooftop antenna, that is the best solution for navigating while moving in a vehicle.

A last comment on accuracy is that under the best conditions, without Selective Availability, locational accuracy is still about 15 m. For most remote sensing projects that degree of accuracy is adequate, but if 15 m is not adequate for some particular use, then differential corrections should be made.

PREPARATION FOR GPS FIELD WORK

Now we get down to practical considerations in preparation for and execution of field work in a remote sensing project. First we will consider things to be done before going to the field, and then items to remember while in the field.

Choice of Maps

Maps are the first item to consider in planning for field work. Whenever possible the map chosen should have a high level of planimetric accuracy and include a latitude and longitude grid at a minimum. Also, it is desirable to have Universal Transverse Mercator (UTM) grid lines or tick marks. It will also be helpful to have a map that provides the UTM zone, geodetic datum, and measures magnetic declination. These items can be acquired from other sources, but having them directly on the map is very convenient.

The choice of map scale depends a great deal on the size of the study area and the amount of detail needed to meet the project objectives regarding map legend and accuracy. Once again, we see that a clear statement of project objectives is vital to all other decisions in the project. For the purpose of this discussion, we will assume the selection of a 1:24,000 or 1:25,000 scale map of the type produced by the U.S. Geological Survey, the Defense Mapping Agency, and the governments of many other countries.

One of the first uses for the map will be to plot a plan for transects that will visit all important data points. Attention can be given to roads and trails that provide the most efficient access to all locations.

Choice of Coordinate System

Most GPS receivers will produce location data in a choice of at least two coordinate systems: latitude–longitude and UTM. There are some advantages to each, but the UTM grid has a strong advantage for personnel moving about in the field. While using the UTM grid, field personnel can determine with each reading the difference from previous readings and see approximately how far they have moved from other sites. Because the UTM grid is based on 1,000-m squares that are printed on recent topographic maps, it is a simple matter to plot positions by hand on a map while in the field. Plotting some of the data points in the field may be useful to check for accuracy when navigating to a set of coordinates that were put into the GPS receiver before going to the field. In this way an instrument operator may pick up errors caused by the instrument, by the terrain, or by the operator. For point plotting it is useful to make a scale marked in meters at the scale of the map being used, and to extend the UTM grid in pencil.

When the locational data collected in the field is to be downloaded later into a computer mapping or GIS program for plotting, field personnel should be aware of the capabilities of the mapping software for converting coordinate systems. Some mapping programs can accept data in either latitude and longitude or UTM, but the latitude and longitude may need to be in a particular form, such as decimal degrees rather than degrees, minutes, and seconds. Mapping programs may also be able to make conversions from latitude and longitude to UTM coordinates, but again may require that the latitude and longitude be in a particular form. Be sure to know the requirements of the mapping software to be used before making a selection of data format in the GPS receiver for locational data collected in the field.

Another important point is that the horizontal geodetic datum of the map must be used as the datum for the GPS receiver. Most topographic maps show the datum information in the margin, for example, North American Datum of 1927. Be sure to set up the GPS receiver to the same datum.

Locational Accuracy

Another important item to consider before going to the field is the level of accuracy needed in locational data. Factors to consider are the area of a sample site, that is, the size of a pixel and the number of pixels per sample, as discussed in Chapter 2. In no case is it advisable to try to locate to a single pixel. Keep in mind that the uncorrected GPS location data may have an error of as much as 15 m, and the size of a sample site should allow for at least twice that amount, as suggested in Chapter 2. For many remote sensing projects that cover large areas of natural vegetation or agriculture, it should be suitable to use a GPS receiver for location of sample sites without making differential correction. As the terrain cover becomes more complex (e.g., in urban areas), greater accuracy is required, and field personnel should plan on using differential corrections. Remember that there is usually a small error in georeferencing of a fraction of a pixel, and this will add to the potential for error in actual field location.

The considerations discussed above pertain to finding the coordinates of a site chosen in the field. When navigating to a known site chosen in the office, either from a map or a georeferenced image, the options are more limited. One can navigate to a site with a single receiver and feel certain of being within 15 m of the true site. To improve accuracy in navigating to a point requires real-time differential corrections and communication by radio between a fixed GPS and a rover. This is a more complex operation requiring more expensive equipment and would be unwarranted in most remote sensing projects.

Selecting an Optimum Time for Field Work

An important part of planning field work is selecting a time when NAVSTAR satellites are in a suitable array to provide optimal accuracy. As noted earlier, there is a useful U.S. Naval Observatory website in which satellite positions for any date, hour, and location can be determined. Information given includes azimuth angle and elevation angle for each satellite with elevation angles above the selected mask angle, also called cutoff angle. The default mask angle is 10° above the horizon. A 10° mask angle is suitable for planning purposes, but 15° or 20° is recommended for better location accuracy in the field by eliminating satellites near the horizon. The information on the Naval Observatory website can be used to create diagrams similar to Figure 3.1 and select optimum times for field work. A blank azimuthal diagram is provided in Appendix 2 and may be used for plotting satellites and potential signal obstructions as well.

Establishing Control Points in a Field Area

It is useful to determine the coordinates of some known control points along with sample sites before going to the field area. Geodetic control points and benchmarks are possible points that may have coordinate data available. If known control points cannot be found, steps can be taken to establish such points, though at a lower level of accuracy than the first order of government control data. Before going to the field, coordinates of easily identified landmarks may be measured from a map and used for an accuracy check in the field.

Georeferenced Image

Before going to the field it would be a great aid to create a georeferenced image of the field area. Also, the image should have an overlay of map features that may not easily be seen on the image alone (e.g., roads, control points, wells, or other landmarks) but would help navigation in the field.

The georeferenced image may be used in two ways. On one hand, the image may be used to select sites to be visited in the field. Selected sites can be marked on the computer screen and a listing of coordinates printed. These coordinates may then be read into the GPS receiver for later navigation to the selected sites. On the other hand, the georeferenced image may be used in the field to select sites that were overlooked in the previous selection of sites for field visitation. Field personnel may prefer to select all sites in the field with a georeferenced image and appropriate overlay information in hand. If an unsupervised classification approach is being used, then an appropriate image to take to the field is a georeferenced cluster map.

An excellent approach to finding field locations is to load a georeferenced image, classification map, or planimetric map into a portable computer that is attached to a GPS instrument. This allows field personnel to display their area and to see a cursor showing real-time locations. This approach is beneficial for finding preselected sites as well as for selecting sites while in the field. Most software allows point coordinates on a map to be downloaded to a GPS as well as from a GPS to map.

Site Characteristics

It is often useful to record features that may affect accuracy at the site where a GPS location is recorded. Plan for collecting this information by identifying in advance what the field personnel should look for while taking a GPS location. Such things as sky condition, tree canopy, nearness of possible obstructions, presence of features that may cause multiple signal paths to reach the GPS receiver (e.g., landforms and buildings), and nearness of electrical power lines or radio transmitters. Obstructions may be plotted directly on a satellite geometry diagram (e.g., Figure 3.1) to determine whether a feature is actually obstructing any satellite.

Summary of Field Survey Planning

In the field work phase of a remote sensing project, planning for the effective and efficient use of GPS is very important. The following checklist may be followed in preparation for GPS work in the field for a remote sensing project:

1. Select maps.
2. Select the coordinate system.
3. Determine the level of accuracy needed for locations.
4. Investigate satellite visibility for the dates of an expected field trip and select optimum times.
5. Determine the location of control points of known coordinates.
6. Create a georeferenced image with an overlay of important points and useful landmarks.
7. Select points to visit (for measurement, observation, or accuracy assessment), and obtain coordinates from the georeferenced image.
8. Identify observations to be made regarding site characteristics.

PROCEDURES IN THE FIELD WITH GPS

If appropriate preparation has been made prior to the field trip, location work with a GPS in the field will proceed with greater effectiveness and efficiency. On arriving for work in the field, there are several guidelines for GPS operations to keep in mind while navigating to known points or finding coordinates for unknown points. Information on what data to collect in support of image analysis is covered in Chapters 5–8.

Awareness of Error Sources

While in the field a GPS operator should continually be aware of features and procedures that may affect the accuracy of the GPS computation of location. The following awareness checkpoints should be kept in mind.

Always hold the GPS receiver with the antenna oriented upward in a position that is unobstructed by the operator's own body. To keep the antenna in a proper orientation, consider mounting an external antenna on the roof of the field vehicle, if the instrument will accommodate it. The small external antenna shown with the instrument in Figure 3.3 is designed to be affixed to the roof of a vehicle magnetically. If moving on foot, consider mounting an external antenna on the end of a 6-foot pole that can be carried properly oriented while walking. This is especially useful when collecting data in tall vegetation such as corn. A pointed end on the pole would allow it to be stuck in the ground while recording information at a sample site. I know of a field person who attached a small antenna to her hat. This allowed her to keep the GPS receiver in her pocket until needed and freed her hands for other tasks.

At each site be aware of any buildings, cliffs, or trees that may affect accuracy by obstructing the view to some of the satellite array and be aware of electrical lines or radio towers that also may interfere with the signal from the satellites.

Initializing the GPS Receiver

If the GPS receiver is an older one that consumes energy rather fast, be sure to start the day with fresh batteries. Newer instruments have improved efficiency, and their batteries last much longer. Frequent use of the GPS instrument before going to the field will familiarize one with its rate of energy con-

FIGURE 3.3. Hand-held GPS receiver suitable for field work with remote antenna attached. Photograph courtesy of Sue McCoy.

sumption. Above all, be sure to carry a good supply of extra new batteries for the GPS receiver. The loss of power can ruin an entire field trip!

At the day's first start-up, a GPS instrument must determine its current location. If the receiver is near the location where it was last operated, initialization may occur quickly. To speed up the initialization time, the operator may put in a pair of coordinates from the map for that day. The quickest way to do this is to use the latitude and longitude values in one corner of the map as input for location. This means that, even though UTM coordinates may be used for all locational work through the day, the initialization is done using latitude and longitude. After initialization, the set-up options may be changed to UTM coordinates. As long as the device is given the correct UTM zone when changing coordinate systems, the results will be correct.

Visiting Control Sites

Consider starting each day at one of the control sites identified before coming to the field to verify accuracy estimates on the GPS receiver. At this time note the information on the instrument concerning satellite status. Most GPS instruments provide a plot of currently visible satellite locations in the sky. This plot should be very close to the plot made with almanac data obtained from the U.S. Naval Observatory website, assuming that the date and time of the almanac coincide with the current date and time. Visit control sites through the day, and occasionally revisit other previously taken GPS locations. This frequent checking and rechecking of coordinates will provide a strong sense of locational accuracy and inform the operator of a likely margin of error in locating pixel locations chosen before coming to the field.

Navigating to Positions

Before going to the field it is helpful to practice navigating to known positions using the GPS receiver. It takes a little practice to see how the receiver performs as it approaches one of its known locations (waypoints) on the various navigation screens. Distance to waypoint will reach zero and then begin increasing, and direction to waypoint will abruptly begin to change direction and appear uncertain as to the bearing. Some receivers will give an audible or visible signal when a waypoint is reached. When the instrument indicates arrival at a control point, the operator's actual location will not always coincide with the control point location. Notice the difference in distance between the site the instrument identifies and the actual location of the site. Practice will build confidence that one is in proximity to, although not necessarily precisely on, the desired location.

If sample points are on a straight-line traverse, it is helpful to supplement the GPS receiver with a bearing on a distant landmark taken by compass or measured from a map. This aid allows one to stay on a line by sight and to rely on the GPS instrument primarily for distance. This would keep any locational error in one direction along the line of sight and minimize error lateral to the line of sight. This approach is only possible when terrain and maneuverability are favorable.

GPS Note Taking in the Field

Field notes concerning GPS surveys may range in detail from a simple list of field sites with observations on possible obstacles to GPS signals to a complete field sheet of information for each field site. When in doubt, go for too much information rather than too little. Remember that it may not be possible to return to the field for additional information.

A place for GPS information may be included on the sheet for other field notes concerning observations or measurements of the phenomenon in question. A complete GPS information sheet should include space for three types of information:

1. Site information: site ID, time of day and date, nearby landmarks.
2. Map information: map datum, UTM zone, and magnetic declination.
3. Environmental information: sky cover, weather, tree canopy, possible obstructions and electrical interference. A circular diagram may be included for marking azimuth and elevation of possible obstruction.

See Appendix 2 for an example of a GPS field note form.

Summary of Field Procedures

1. Be sure the GPS antenna is properly oriented and unobstructed at all times.
2. Initialize the instrument using nearby latitude and longitude coordinates from a topographic map corner. After initialization, switch to other coordinate systems if desired.

3. Check the satellite almanac with the instrument plot of visible satellites for the time and day.
4. Visit known control sites if possible to check for accuracy. Derive coordinates from a map if known coordinates are not available.
5. Navigate to sample sites.
6. Make appropriate notation of features that may affect GPS locational accuracy.

4

Field Spectroscopy

The primary purpose of this discussion of field spectroscopy is to provide a guide for collecting field reflectance spectra. In order to understand the basis for field procedures, some background in the characteristics of electromagnetic radiation reflectance geometry is needed. For those who are unacquainted with this literature, there is a brief discussion of radiation reflectance with some appropriate citations for further reading. Spectral data analysis is not included in this discussion, but references to that topic are provided.

FUNDAMENTALS OF REFLECTANCE

Procedures for collecting reflectance spectra in the field are based on the behavior of reflected electromagnetic radiation. Field personnel, therefore, must have a clear idea of radiation sources, the basic geometry of the radiation environment, and an understanding of the interactions which occur when solar radiation encounters an object.

The geometry in its simplified view is shown in Figure 4.1. The diagram

42

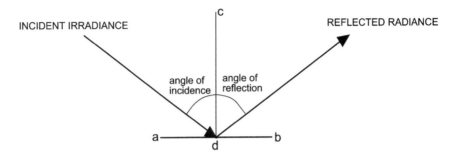

FIGURE 4.1. Reflectance geometry of direct solar irradiance assuming specular reflectance from a single source. The angles of incidence and reflection are equal angles measured from a line (*cd*) perpendicular to the surface. Line *cd* is not a zenith line unless the surface is horizontal. Line *ab* is the orientation of the incidence–reflection plane measured as a horizontal angle (azimuth) from a given reference, usually true north.

depicts radiation geometry in the case of specular, or mirror-like, reflection. In that type of reflection we see solar radiation hitting the surface at the incident angle, which is measured from the zenith angle. The angle of reflection is also measured as a zenith angle and is equal to the angle of incidence. The zenith angle is measured from a perpendicular to the target surface so that directly overhead, on a horizontal surface, is 0°. The horizontal angle, or azimuth, is usually measured as an arc of a clockwise turning angle with north at 0° and east at 90°. The azimuth defines the orientation of the planes of incidence and reflection.

The real-world environment is illuminated in a more complex way than shown in Figure 4.1. Direct solar radiation plus scattered radiation (i.e., skylight) form a great hemisphere of incident radiation hitting the surface. Likewise, the reflected radiation is typically diffuse rather than specular, and forms a hemisphere of reflected radiation. Hence, the total radiation environment is often referred to as bihemispherical and consists of incident hemispheres of radiation and a diffusely reflected hemisphere.

Direct solar radiation is by far the dominant source, contributing as much as 90% of the total incident energy in a perfectly clear sky (Figure 4.2). Scattered radiation accounts for 10–20% of total energy over the spectrum, with the greatest amounts contributed by the short wavelengths, ultraviolet and blue. Reflections from nearby objects such as trees, buildings, clouds, or even the person operating the instrument comprise yet additional sources of inci-

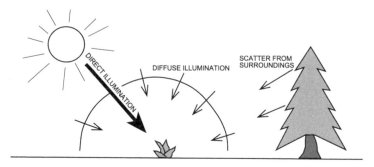

FIGURE 4.2. On a clear day direct illumination is dominant. Operators need to be aware of the potential for strong illumination from nearby objects. Adapted from Curtiss and Goetz (1994). Used with permission.

dent radiation. Although all these radiation sources are part of the total, in practice only the direct solar radiation is considered. However, radiation from nearby objects must be minimized when collecting data. In order to satisfy an underlying assumption that spectral data collected in the field is primarily from a single source (i.e., direct solar radiation), it is best to collect reflectance under a perfectly clear sky. Of course, field work cannot always be done under these ideal conditions, and possible adjustments in field procedures under heavy haze or cloudy skies will be discussed.

Bidirectional Reflectance Factor

Consider the incident solar energy (irradiance) and the reflected energy (radiance) as two elongated cones, each forming small solid angles at a point on the target surface (Figure 4.3). The reflectance of that particular point on the target can be expressed as a ratio of the radiance to the irradiance.

In order to fully express the hemispherical reflection characteristics of an entire target, rather than only a point, it would be necessary to measure the irradiance and radiance at all possible sensor positions and all possible source positions. This hemisphere of reflected radiation within a specified spectral range is referred to as the target's spectral indicatrix (Goel, 1988; Curran, Foody, Kondratyev, Kozoderov, & Fedchenko, 1990). If one were to devise a 1° sampling grid over the entire reflectance hemisphere of a target, 400 million measurements of all possible irradiances and radiances would be needed to characterize the spectral indicatrix of that target. The result of this effort would produce a bidirectional reflectance distribution function (BRDF). An indicatrix in the form of a polar diagram and plot for a shrub is shown in Figure 4.4. Even though the indicatrix shown would be based on relatively few measurements, it is sufficient to show that target radiance has definite directional bias. Clearly, this is not a procedure anyone would want to follow for a great many targets in a remote sensing project.

The practical alternative to measuring the indicatrix is to measure samples of radiance from the target along with the radiance of a standard white reflectance panel to represent irradiance. Reflectance (r), then, is a ratio of target radiance to panel radiance.

$$r = (\text{radiance of target/radiance of panel}) \, k$$

The constant, k, a panel correction factor, is a ratio of solar irradiance to panel radiance and ideally should be near 1. We assume that the reference panel is a Lambertian reflector; hence, it would have the same correction constant regardless of variations in zenith or azimuth angles of the incident radiation (Nicodemus, 1977; Robinson & Biehl, 1979; Jackson, Clarke, & Moran, 1992). A Lambertian surface has a perfect hemisphere of reflection with no directional bias. Like many assumptions, this one is not strictly true, since a perfect Lambertian surface is not attainable.

A common material for reflectance panels is Spectralon®, a hard and durable white unglazed ceramic surface having a reflectance averaging about 98.2%, varying with wavelength from 95.0 to 99.3%. Other materials, such as

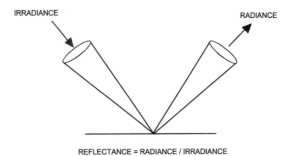

FIGURE 4.3. The cones are the solid angles formed by direct irradiant flux and diffuse radiant flux. They are segments of hemispheres of irradiance and radiance, respectively.

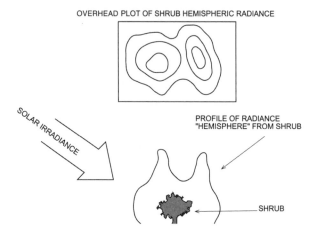

FIGURE 4.4. Hypothetical example of variation in radiance from a shrub. The profile shows the shape of the hemispheric radiance corresponding to the map view above. Data obtained from a spectrometer in this situation would depend on the selected viewing position.

Kodak reflectance paint containing barium sulfate, may be sprayed onto a sheet of aluminum. Also, a Kodak gray card may be used. Gray cards are inexpensive and disposable, but tend to have a directional reflection bias. In a pinch, one could use plain white construction paper, but its reflective properties are the least uniform, compared to the other panel materials. All materials have the problem of collecting dust through the course of the day, and care must be taken to maintain a clean reference panel. A ceramic reflective surface must be cleaned with clean water and very fine waterproof sandpaper. Using such a panel in a dusty environment or with dirty hands may require cleaning the reflective surface several times through the day.

USES OF FIELD SPECTRA IN IMAGE ANALYSIS

Field spectroscopy is the measurement of the interactions of radiant energy with *in situ* objects in the environment. This discussion will focus only on the measurement of reflected solar energy in the visible (V) and infrared, including both near infrared (NIR) and short wavelength infrared (SWIR), wavelengths (400–2,500 nm). See Milton (1987) for further discussion of applications.

Creation of a Spectral Library

One of the primary uses for hyperspectral data is to identify unknown materials by comparing spectral curves obtained in the field or from hyperspectral imagery with spectral curves of known substances. Several extensive spectral curve libraries are already available, but there are many possibilities for additional spectral curves that could be used to create a spectral library for particular cover materials in a project area. For this purpose, it might be necessary to obtain new spectral curves either in the field or in the laboratory to suit specific needs. For example, it might be necessary in a project to collect library spectra from plants in various stages of stress or other altered target materials in order to identify them on images. Existing spectra for soils, minerals, and vegetation may be found on the websites of the USGS Spectroscopy Laboratory at *http://speclab.cr.usgs.gov/*, and in the Purdue/LARS Vegetation and Soils Field Research Data Summary at *http://shay.ecn.purdue.edu/~frdata/ FRData/*.

Selection of Wavelengths and Seasons

Field spectroscopy can play an important role in identifying optimal wavelengths for sensing particular objects and materials prior to obtaining images of an area. This is particularly useful as airborne and satellite hyperspectral data become increasingly available. In the same way it might be worthwhile to identify times of the year in which similar plant species are best discriminated.

Modeling

One of the more exciting aspects of remote sensing is the potential for modeling the interaction of solar energy with the biophysical matter of the surface environment. If one can determine the relationship between reflected energy and the biophysical character of a surface cover material (e.g., plant community composition and cover density), then by inversion of the model one may measure the reflected energy and predict the identity or condition of the biophysical material. One of the earliest examples is Miller and Pearson's (1971) use of a spectrometer connected to a programmable calculator to estimate biomass in a grassland. In order to establish the regression relationship between biomass and spectral response, Miller and Pearson first had to measure spectral reflectance and the corresponding biomass for numerous sample plots in a northeastern Colorado grassland.

Spectral Mixing and Unmixing

Examining the information contained within a pixel in satellite imagery is a specific application of modeling called spectral unmixing. By taking field spectra from each cover type within a mixed pixel, it is possible to create a spectral curve that is a mixture of all the components of the ground pixel area. This synthetic spectral curve should be comparable to a pixel in the satellite data that has recorded the response of the same mixture of cover types. The satellite data can then be unmixed to estimate the proportions of each cover type contributing to the spectral response (Adams & Smith, 1986). There are, of course, limitations to this technique. The investigator must know which end members (individual components of ground cover) can be expected in an area, and the number of end members that can be identified is controlled by the number of spectral bands available in the data set.

Spectral Characterization of Biophysical Features

In order to understand fully the character of the interaction of solar radiation with features on the surface, it would be necessary to determine a complete spectral indicatrix of the object, as described earlier. It is sufficient to know that most surfaces have a directional bias in their reflection pattern. Study of the radiance diagrams (e.g., Figure 4.4) will show the importance of maintaining a consistent geometry in the positions of the sun, the target, and the field spectrometer. The diagrams show that reflection varies with viewing angle. Therefore, the field investigator must constantly be aware of the angles between the spectrometer, the target, and the sun. This will be discussed more fully under the topic of field procedures.

Spectral characterization also refers to the relation of wavelengths and intensities of reflection of selected biophysical targets. Field spectroscopy may be used to study the spectral differences among the various cover materials on the surface, even to the subtle differences among species of plants. In this way one may anticipate the degree to which surface features might be differentiated by satellite or airborne sensors. It can become clear in advance which cover types might be easily distinguished and by which spectral bands. Likewise, it is possible to determine which cover types will be difficult to identify spectrally. Advance knowledge of this type is particularly useful when deciding how spectral classes would be grouped into information classes when identifying training sites in a supervised classification scheme.

ASSUMPTIONS WHEN MEASURING REFLECTANCE SPECTRA

Certain geometric and environmental conditions are assumed to exist when applying field spectroscopy, and a spectrometer operator must make every effort to remain aware of these assumptions. Refer to Curtiss and Goetz (1994), Milton (1987), Salisbury (1998), and Robinson and Biehl (1979) for further discussion of these concepts.

Assumption 1: The FOV of the Sensor Is Known

As with most sensors, the spectrometer's FOV in combination with the distance to the target determines the area that is sensed. Figure 4.5 shows the FOV geometry by which one can determine the optimum distance from instrument to target. The tangent of (θ), one-half the angle of the instrument's field of view, multiplied by the distance (d), gives the radius (r) of the field of view at the target:

$$r = d \tan \theta$$

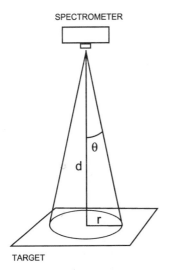

SPECTROMETER

TARGET

FIGURE 4.5. Geometry of the field of view (FOV). The diameter of the base of the cone at the distance, d, must be computed.

In order to hold the instrument at a distance that will satisfy the next two assumptions, the FOV must be known.

Assumption 2: The Reference Panel Fills the FOV

This assumption must be met in order to measure a spectral curve that is representative of the reference panel. To avoid an influence of spectral radiance from adjacent materials, the field of view diameter should fill no more than one-half of the reference panel. The operator can control this by adjusting the distance from the spectrometer to the reference panel.

Assumption 3: The Target Fills the FOV

Exactly what fills the FOV depends, of course, on the target. However, anything that is not a part of the chosen target must be outside the field of view. As above, the target area should be at least twice the diameter of the field of view. The distance to the target scene may be adjusted to obtain a bigger or smaller FOV, as needed.

Assumption 4: Irradiance Is Constant While Measuring Both the Reference Panel and the Target

Even if the sun is shining in a clear sky, it is possible to have small differences in solar irradiation due to thin cirrus clouds not visible to the eye or due to movement of dust through the atmosphere. These minor effects cannot be controlled. Scattered clouds are the most likely feature to cause a change in illumination in a short period of time. This effect is minimized if the time required for the instrument to take a measurement is very short and if the operator can redirect the instrument quickly from reference panel to target. If illumination conditions are very stable, one may take fewer reference panel measurements and interpolate between selected reference panel measurements.

Assumption 5: Direct Solar Irradiance Is the Dominant Source of Energy Incident on the Target

This assumption is never fully met because skylight is always present, even though ignored. As mentioned earlier, direct solar irradiance accounts for up to 90% of the total on a clear day, but as atmospheric haze increases, the proportion of scattered skylight increases. Also the short-term variability increases, making the satisfaction of assumption 4 more difficult.

Assumption 6: The Sensor Has a Linear Response to Changes in Radiation

There is little the operator can do other than to make sure that the manufacturer of the instrument has provided such a fundamental quality.

Assumption 7: The Reflectances of the Standard Panel at Various Wavelengths Are Known, and They Do Not Change during the Course of Work

For computation of relative reflectances among targets, one can get by without knowing the spectral reflectance of the panel. It is crucial though that the panel reflectance not change due to dirt or moisture during the period of measurement. To protect the integrity of measurements, it may be necessary to clean or change the reflectance panel periodically. Computation of absolute reflectance, which would be used for calibration purposes, requires spectral reflectance of the panel for accurate corrections.

FIELD PROCEDURES FOR MEASURING REFLECTANCE SPECTRA

Consideration of the assumptions outlined above makes us realize that there are several geometric and environmental variables present, with potential for error in each. Therefore, the spectrometer operator must strive to minimize error with consistent procedures for each measurement. Only through consistent procedures can data from one time be compared with data from another time of measurement. To give some idea of the points where consistency is most important, the following field procedures are suggested.

1. Maintain a consistent viewing geometry relative to the solar azimuth angle. This procedure assures that spectral samples are taken from the same portion of the target radiance hemisphere with each measurement. The operator must change the viewing azimuth through the course of the day as the solar azimuth angle changes. This is easy to do by standing with one's back to the sun at every measurement site while measuring both the reference panel and the target. Of course, this position must be assumed in a way that does

not cast the operator's shadow on the reference panel or the target. If the sensor aperture is at the end of a fiber optic cable, as shown in Figure 4.6, this is easy to do. Having the operator's back to the sun also has the advantage of minimizing reflected incident radiation from the operator's body. Consistency in distance of the sensor from the target must also be maintained as a part of the viewing geometry.

If a fiber optic probe is not available on the sensor, it is good to mount the sensor on a tripod. Handheld sensors without fiber optics usually have greater variation in viewing geometry, plus greater chance of reflected radiation due to the proximity of the operator. Care must be taken to avoid casting shadows into the target area from the operator, bystanders, a tripod, or a truck boom. In some instances it is necessary to use an overhead platform (ladder or cherrypicker) in order to expand the FOV sufficiently to view a representative scene. In a row crop, for example, an integrated spectral response of soil, shadow, and crop may not be obtainable for an operator standing on the ground. Therefore, height above the target may make a difference in the reflectance recorded by the spectrometer (Daughtry, Vanderbilt, & Pollara, 1982).

FIGURE 4.6. Field operator taking spectra of rock outcrop, using back-mounted spectrometer with fiber optic cable connected to probe. Photograph courtesy of Analytical Spectral Devices, Inc.

2. Determine that the reference panel and the target each overfill the FOV of the sensor while taking a measurement. This is a condition that must be considered in order to establish an appropriate distance from the sensor to either the panel or the target.

3. Vehicles and persons other than the operator should be kept several meters away from the target. The operator should wear dark clothing and, as mentioned above, stand facing away from the sun. Tripods or truck-mounted booms should be painted flat black. These precautions will reduce the variability of measurements.

4. Multiple measurements should be taken when measuring vegetation foliage in order to compensate for movement of the target by wind. Gusting wind is known to cause large variations in spectral measurements of vegetation, and even light breezes can cause variations on the order of 12%. Although many researchers simply average many spectra, Lord, Desjardins, and Dube (1985) found that the mean reflectance factor overestimates the true value because of extreme changes during gusts of wind. They suggest that the median value as a wind correction provides a result closer to the true value. Forty scans over a short period should produce good signal-to-noise relations for any wind-disturbed target.

The effect of wind on water also presents a problem in getting reliable reflectance spectra. Disturbance of the water creates many little specular surfaces, resulting in glitter with a poor signal-to-noise ratio. Averaging many spectra is a solution to this problem. If the glitter is caused by helicopter downwash, the operator should determine a flight height that does not disturb the water surface.

5. Measurements should be made at a time when direct solar flux is the dominant incident radiation. Ideally, there should be a perfectly clear sky with no haze, but in the real world this is an elusive condition. Most of the time compromises must be made. When there are many small cumulus clouds moving rapidly across the sky, both magnitude and frequency of variations in incident radiation are greatest. This increases the potential for a difference in illumination between measurements of the reference panel and the target, and necessitates great effort to minimize the time between their measurement.

Often the day chosen for field work turns out to be overcast with clouds. Although an overcast sky may look uniform in illumination, there are actually great variations in thickness of the clouds, hence variations in brightness. The result is similar to the variations in movement of scattered clouds across the sky. Furthermore, the incident light under an overcast sky is all scattered light and has a different spectral character than direct sunlight. The results from measurements under overcast conditions have high variability and may not compare with measurements taken at any other time.

Atmospheric haze and thin cirrus clouds are often overlooked as possible sources of error in field spectral data. The sun seems to be bright and shadows exist, though perhaps they are diffuse. Figure 4.7 shows that the solar spectrum under hazy conditions has the same spectral distribution as the clear sky, but at a much lower magnitude.

This difference becomes especially important in the SWIR where there is lower energy to begin with. The difference between spectra collected in the high sun of summer and the low sun of winter has the same characteristic, and a graph of seasonal spectra would look nearly the same as the clear versus hazy sky spectra. This should not prevent work from proceeding, but the operator needs to be aware of the importance of sky conditions when analyzing spectral data, especially if the data cover several dates.

A solution to the frustration of highly variable illumination is to use an artificial light source either in the field or in the laboratory. Some researchers have brought samples of targets back to the lab as a regular practice simply to control for the high variability of natural illumination. When vegetation is involved, it is necessary to make the measurements as soon as possible and to pack the samples in ice to minimize wilting. For some purposes artificial illumination is adequate, but solar illumination is required for projects involving field-to-image calibration simultaneous with sensor overflights.

The question of direct versus indirect solar illumination applies also to targets in shadow. Remember that skylight has a spectral distribution strong in the blue and greatly reduced in the near infrared relative to direct sunlight,

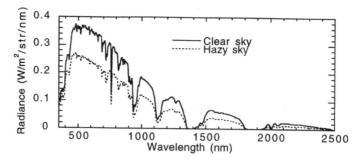

FIGURE 4.7. Solar radiance under clear and hazy skies. In addition to attenuation of energy, as shown, hazy sky causes solar irradiance to be highly variable. From Salisbury (1998). Used with permission.

and the longer end of the SWIR may be essentially missing. This will make data from targets in shadow different from data for the same materials in direct sunlight.

The important procedure to remember is that the greater the atmospheric variability, the more frequently an operator should record reference panel spectra immediately before or after each target spectra. Under very stable atmospheric illumination conditions, reference spectra may be reduced to a frequency of 6–8 minutes and applied to a number of target spectra taken during that time. If the reflectance resulting from the ratio of the new reference panel reading and the previous panel reading is anything other than 100%, then either solar irradiance or the instrument responsivity has changed. Depending on the warm-up time of the instrument, it may be necessary to recheck the reference panel frequently until the reflectance curve appears to remain stable between readings. Remember that the most frequent error made by field spectrometer operators is to acquire too few reference panel readings. This is especially true in overcast or other variable sky conditions.

6. Field work should be limited to periods of high sun. Great variation in incident radiation occurs as sunlight passes through longer stretches of atmosphere with increasing path lengths through haze and dust. It is best to restrict measurements to a period about 2 hours before and after solar noon. In a very clear atmosphere it is possible to extend the time up to 30 minutes each way. Such a constraint reduces the length of the effective workday and may extend the number of days needed in the field. This is another reason some researchers take samples to the lab for measurement under artificial light. Figure 4.8 shows the effect of low sun angle on the spectral curve of kaolinite. The great increase of noise in the curve made at lower sun angle obscures the characteristic absorption band of kaolinite.

USE OF ARTIFICIAL ILLUMINATION

Although it is usually desirable to obtain field spectra under direct solar illumination, when comparing with imagery there are instances when it is advantageous to make spectral measurements under artificial light.

1. The use of artificial light allows ideal control of viewing and illumination geometry.
2. With artificial light, measurements can be made in the absence of direct solar light, including cloudy weather and nighttime.

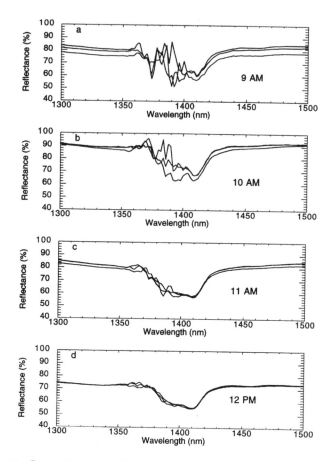

FIGURE 4.8. Spectral scans of the mineral kaolinite taken at four different times (a, b, c, and d) through the morning show the effect of zenith angle on detectability of features in spectral response curves. Kaolinite's response character in the area of the water absorption band (1400 nm) becomes more apparent as the sun gets higher. From Salisbury (1998). Used with permission.

3. Under the controlled conditions of artificial light, wind and haze are not problems.

A disadvantage of measuring spectra under artificial light is that the validity of calibration or comparison of the sample data with airborne or satellite image data is less certain. Of course, some materials cannot be brought to the laboratory because of size or susceptibility to rapid change with time.

When making spectral measurements in the lab, quartz halogen lamps of 200–500 watts power with a spectral temperature of 3,400° Kelvin are recommended to provide a light that is spectrally similar to sunlight (Curtiss & Goetz, 1994). Incandescent lights are deficient in blue wavelengths, and fluorescent lights are too strong in blue.

Care should be taken to maintain sufficient distance between lights and samples to prevent heating of the sample, especially when measuring vegetation spectra. Some manufacturers of field spectrometers now offer various instruments with a built-in light source for measuring spectra with artificial light.

A battery-powered hand-held light probe is useful for making spectral measurements with artificial light both in the field and in the laboratory. A fiber optic cable connects the probe to the spectrometer (Figure 4.9).

FIELD NOTES FOR SPECTRAL SURVEYS

Appendix 2 covers general guidelines for taking field notes in most types of remote sensing projects, and provides forms that could be used for doing so. However, the needs for field notes in a field spectroscopy project are specialized, and some elements of field notes for spectroscopy are listed below. In addition to notes for date and time of each scan, the spectrometer should be able to time-stamp each scan for reference and identification.

FIGURE 4.9. Light probe being used to obtain spectra of rock cores under artificial light in laboratory. Photograph courtesy of Analytical Spectral Devices, Inc.

The main thing to keep in mind is that there are many variables affecting spectral responses, and no one can remember all the physical and instrumental details associated with each scan. Also, the instrument operator may not be always be the person who does the data analysis. Therefore, notes and photos should be taken on the assumption that someone other than the operator may analyze the spectral data for the project.

A camera is an excellent way to record many details of the target, set-up geometry, terrain, and sky conditions. Even with pictures, it is necessary to make notes to record additional information in reference to a photograph. A form can be devised to allow the operator to fill in blanks to provide the following basic information:

1. Name of project and site location
2. Date and time
3. Type of target measured
4. Name of operator and others in the field
5. Environmental information on topography, vegetative cover, soil or rock, sky and wind condition, proximity of features that may affect irradiance
6. Reference curve used (new scan or previous scan)
7. Details of geometry including solar zenith angle, solar azimuth angle, instrument azimuth, zenith and distance, and size of field of view.

5

Collecting Thematic Data in the Field

One of the most important steps in any remote sensing project is to make proper field measurements or observations of the phenomenon being studied. This is also the most difficult part of the project for several reasons. First, making field measurements is very time-consuming and often tedious work. But even more difficult is the problem of determining what should be measured, or observed, and what method should be used. A complicating factor in choosing an appropriate method is the difficulty of finding research publications in which the method is described in detail. Also, seeing several methods available for a given measurement, along with some controversy about which method is more valid, often confuses the novice. As in most aspects of field work, validity of measurements is usually a function of the objectives of the project and the user of the end product.

This chapter provides an approach to planning the field work phase of a project, but it does not attempt to provide a method suitable for all the types of measurements and observations that might be made. For that purpose a list of readings organized by topic will introduce field work planners to the literature (see Appendix 1). These references are by no means exhaustive but should provide a starting place for understanding field measurement methods. The text of this chapter is concerned with ideas that should be considered when planning field measurements. Chapters 6, 7, and 8 are concerned with an overview of measurement methods for vegetation, soil and rocks, and water, respectively. These three ground cover materials are components of all land cover/land use types, and they have a significant place in nearly all remote sensing projects.

PRELIMINARY PREPARATION

Before beginning actual measurements of any features in the field, several steps described in earlier chapters must have been completed.

1. Project objectives must be clearly established, as they will determine what features to measure and in what level of detail. Other steps follow from careful planning of the objectives.

2. A classification scheme for all surface categories must have been selected. By this time it should be known how vegetation communities will be mapped. Classification may be at a very general community level with broad categories such as forest and rangeland. It may be a physiognomic classification with classes such as broadleaf deciduous trees, needleleaf evergreen trees, grass, and shrubs. A more detailed classification based on floristic composition would include a combination of physiognomic classes with common names of community dominant plants at the generic level such as pine forest, oak–hickory forest, bluestem grassland. The classification scheme chosen should be the one that fills the needs of the project.

REFERENCE INFORMATION

Reference materials include all sources of information other than the image data itself, that is, field observations and measurements, aerial photographs, maps, and other archival materials. The term "ground truth" was long ago replaced by the more appropriate term "reference information" to be more inclusive than "ground" and less absolute than "truth." Although no project should be conducted without visits to the field area, it is often possible to select, with great care, training sites and accuracy assessment sites from an aerial photograph or possibly a suitable thematic map showing land cover types.

Aerial Photography

Before beginning a project, field personnel should determine what aerial photo coverage is available, and a preliminary interpretation could be completed before going to the field. Examination of photos before going to the field helps to identify sites to visit for identification of cover material, identify possible ground control points, and provide a base of information for later update. Aerial photograph updates are especially needed in agricultural or urban fringe areas where change may be frequent. If available, aerial photographs taken in both the growing season and in the dormant seasons are very useful. Color infrared photography provides valuable information and should be used when available. Otherwise, black-and-white photography also serves. Contact the appropriate government agency for photo coverage in the area of interest. In the United States, the U.S. Department of Agriculture Aerial Photography Field Office, Salt Lake City, Utah, sells aerial photography for the U.S. Forest Service, the Bureau of Land Management, the USGS, and interagency photography such as the ongoing National Aerial Photography Project (NAPP). Ordering information and a film catalog are available on their website, *http://www.apfo.usda.gov/*. Orders are best initiated by phone at 801-975-3503. Information about NAPP and other coverage can be found on a USGS website at *http://edcsns17.cr.usgs.gov/ finder/finder_main.pl?dataset_name=NAPP.*

Aerial photographs should be used for training site selection only if they are close to the image in date, or if it can be determined that cover materials have not changed over time. Visits to the field should be made to verify photo interpretation done in the laboratory. If both the image and the photo are from an earlier time, the cover types in the field may be different from both, especially in agricultural areas or urban fringes. In these cases, field work will be of limited value, and the primary reference material will be aerial photographs.

Other sources for photographs and imagery include many state websites where the public can download photo products, such as orthophotos. Also, the Eros Data Center in Sioux Falls, South Dakota, is a primary source of imagery.

Maps

Thematic maps of cover type should not be used for training or accuracy site selection unless there is no alternative. Maps are generalized information and likely are based on different definitions of categories as well as a different minimum cell size than is being used in a particular project. Either of these deficiencies makes most maps unreliable and invalid for use as reference materials, other than for general knowledge of the area.

Experiments can be made by collecting subsets of data or observations from the field and comparing results with comparable data collected from a map or photograph. If there is an unacceptable difference in several mapping categories, it is best to rely on data collected from the source that is closest to the image in time. Usually this comparison will favor data from the field. Some researchers have found that even when photographs are recent, data derived from them are not as accurate as data obtained on site. If photographs of a recent date are used for identifying plant communities by dominant species, or for observing types of urban land use, their reliability is usually high. If observations require more detail, such as cover density or crown closure, field data are more likely to provide the better result.

While aerial photographs are valuable tools, it is important to be aware of their limitations. Experiments to determine the relation between data obtained in the field and data from photo interpretation will help reveal the limitations. For example, photo estimates of crown cover in a wooded area could compare well with the same values collected in the field. However, photo estimates of cover in grasslands may vary greatly with field data. A few test sites will show what may be estimated reliably from an aerial photograph and which data must come from field measurements.

Metadata

The number of geospatial databases is growing rapidly to include not only imagery resources but also maps and other data that would have potential use as reference information in remote sensing projects. This rapid growth has led to the development of metadata as a means for cataloging and describing the nature of available data. Metadata is online information about data sets, and may be used like the catalog in a library. For example, metadata might consist of imagery dates, quality assessment, cloud cover measurement, and geographic coordinates. It can be a great aid in selecting imagery or reference data. Because geospatial data sets are of particular interest to remote sensing projects, Appendix 3 has been put together and provides a list of metadata sources.

TYPES OF FIELD WORK

Two types of information are collected in the field: observations and measurements. A project may use only one of these types or a combination, depending on the project objectives and the level of detail needed in the final product.

Simple Observations

The most common effort in the field is simple identification on the ground, or on photographs, of categories in the proposed map legend. This usually consists of deciding which category of the legend a surface feature best fits. An example would be deciding whether to name a site "grass" or "mixed grass and shrub," assuming that both categories are part of the planned map legend. This means a decision has been made in advance concerning how to differentiate, for example, between "grass" and "mixed grass and shrub." If it is acceptable for the differentiation between "grass" and "mixed grass and shrub" to be approximate, then field personnel may estimate visually whether a given site falls into one or the other map category. The decision on differentiation between "grass" and "mixed grass and shrub" must be based on a rational definition that has meaning to the end user of the final product. Usually this definition would have a break point between the two categories based on estimates of percentage of shrubs present. Other examples might include identification of rock types, land use, agricultural crops, and vegetation types.

Measurements

If the project requires quantitative statements of surface information, then it will be necessary to make measurements of surface phenomena. One example would be to make a quantitative differentiation of vegetation cover density. To do this, the field person would need to make enough vegetation density measurements to represent all the variations in vegetation cover over the study area. The purpose would be to develop cover density classes with quantitative break points (e.g., <10%; 11–40%; >40%), rather than qualitative breaks (low, medium, high). Other examples would be to quantitatively differentiate crown cover, water depth, water turbidity, building density in residential or commercial urban areas, soil moisture content, mineral content of rocks, snow moisture content, crop vigor, crop moisture content, and numerous others.

In many remote sensing projects a single objective (e.g., mapping cover density) would require that all sample sites be measured and analyzed by the same method and at the same level of detail. In other projects only cover type may be needed for each category of the map legend, and no measurements would be required. Another possibility is that certain map categories are of greater interest, and only certain parts of the map legend may have precisely defined subcategories that must be measured. Then other map categories are of less interest and require only identification by observation. An example of this latter situation would be mapping rangeland cover density by measurements and showing other areas by general undifferentiated categories such as agricultural land, urban, water.

FIELD WORK TASKS

Following the maxim that methods are usually tied to objectives, we should begin by examining the likely objectives of field measurement and observation in a remote sensing project.

Compiling Training Sites

In a supervised classification approach to mapping, one of the primary steps is visiting field sites for the purpose of establishing training sites from which training data are derived.

Identifying Cluster Composition

In an unsupervised classification the field person goes into the field with a cluster map for the purpose of matching mapped clusters with features on the ground. At this stage he or she is concerned with the relationship between spectral classes (the clusters) and map categories (the legend). Because certain clusters may represent spectral differences within the same ground cover, the field person must decide how best to group clusters. This grouping should be done with knowledge of the physical differences in ground cover (e.g., cover density variations) represented in each cluster. Depending on the objectives of the project, it may be necessary to group several clusters into only one map category. However, the different clusters in that case would also present an opportunity to subdivide the map categories to more closely match the spectral information in the clusters. If this is done, the field person still has the option of merely identifying qualitatively the differences in each cluster or actually measuring their physical characteristics for a more quantitative definition of map categories.

Identifying Image Features

Although most of the previous discussion relates to producing thematic maps of cover types, many remote sensing projects result in maps that depend more on image feature interpretation than on classification. Such interpretation may be done by purely visual means or by machine enhancement procedures. In these cases, the objective of field work is to verify the image features. This is true of mapping linear features such as faults, escarpments, stream patterns, and linear or circular features of unknown origin as may be used in mineral or petroleum exploration.

Developing Data for Spectral–Biophysical Models

In many projects the objective is only to develop a model that operates on the relationship between the spectral character of a feature and its physical or biological character. Such models are usually either deterministic mathematical models or statistical models based on correlation analysis of physical and spectral variables.

One common example of a spectral–biophysical model is measuring the biomass of rangeland. To determine the relationship between the biophysical properties (e.g., biomass) and spectral response of rangeland, it is necessary to cut and weigh vegetation in a sufficient number of vegetation plots to fully represent the biomass corresponding to a pixel array. Even using shorthand methods that involve weighing a few representative samples and estimating the rest, this is slow work.

Other examples of biophysical to spectral models would be for estimating water depth, water quality, soil moisture, water consumption by plants, crop vigor, crop moisture, runoff coefficients, and urban socioeconomic conditions related to building density, amount of pavement, and so on.

SELECTION OF PHYSICAL FEATURES FOR OBSERVATION OR MEASUREMENT

Perhaps the toughest question for a remote sensing field person is what to measure in the field. There are countless possibilities for both direct and surrogate measurements, but practical limits (budget and time constraints) dictate how much time can be spent in the field. In addition, many possible measurements would have only a weak association with variations in the spectral response of a particular feature. It is important to measure variables that show a strong relationship to spectral response. Spectral response, therefore, is the key to selection of field measurements.

Spectral Response of Surface Materials

It is important to have knowledge of the spectral response of several general materials, such as vegetation, soil, water, concrete and asphalt, across the reflectance spectrum from 350 to 2,500 nm. Further knowledge of reflectance spectra should include variations caused by drought and senescence of vegetation, moisture in soil, and turbidity in water. Beyond this level, one should begin to develop a knowledge of variations in spectra due to species differences, organic matter in soil, algae in water, and so forth. In short, there is a great deal of benefit to having familiarity with the spectra of the materials being mapped. An excellent resource for learning some of the details of spectral response is the book by Swain and Davis (1978).

Spectral libraries are available from the Laboratory for Applications of Remote Sensing (LARS) and the USGS websites. LARS has over 200,000 spectra of soils and vegetation that can be downloaded from their website at *http://shay.ecn.purdue.edu/~frdata/FRData*. The USGS Spectroscopy Laboratory website (*http://speclab.cr.usgs.gov*) has a valuable library of spectral response curves of minerals and vegetation that can be downloaded. Hard copy is also available (Clark, Swayze, Gallagher, King, & Calvin, 1993).

Knowledge of spectral responses helps in selecting appropriate variables for measurement in the field by relating the spectra to the features that control variations in spectral response. A list of variables controlling spectral response is also a list of measurements and observations that could be made in the field. The list of field variables may be divided into those that can be determined by observation alone and those that must be measured. Observations, for example, would include spectral response controls such as plant species and general cover type (vegetation, bare soil, water, concrete, etc.).

Measurements would be needed to determine variations in vegetation cover density, moisture content, building density, biomass, turbidity, and water depth.

Keep in mind that measurements of physical features relative to spectral response are often redundant. For example, vegetation cover density, leaf area index (LAI), biomass, and crop height are each related to the effect of total volume of vegetation on spectral response. Reflectance spectra due to vegetation cover density is, of course, less sensitive to total vegetation once the plant cover has reached 100%, but reflectance continues to vary with increasing biomass and LAI beyond 100% cover. Once again, the statement of objectives and the needs of the end user will determine the best choice of measurements and avoid unnecessary, redundant measurements that can greatly increase time in the field with no gain in mappable information. Let a careful statement of project objectives made at the beginning determine which physical variables to measure or observe.

One guide for which variable to measure is to select those that provide the needed information and are most readily measured. Sometimes it is helpful to find, through the literature or by experimentation, a surrogate. For example, plant height may in certain instances be adequately correlated with biomass so that height, or crown diameter, may be measured. In shrubby vegetation, it is possible to select a representative branch of a shrub, weigh it, and estimate how many such branches make up a particular shrub being sampled. Consult the published literature for procedures appropriate to the project's particular needs. Vegetation measurement methods will be discussed further in Chapter 6.

MEASUREMENT PROCEDURES IN THE FIELD

For every physical phenomenon to be measured, a body of literature exists on how to make the measurement. Not surprisingly, no single method is universally accepted in most cases, and often there is some controversy concerning which is appropriate.

Whichever method is chosen for making measurements, it is the responsibility of the researcher to select methods that are acceptable to professionals in the subject area in question. If measuring rangeland vegetation, for example, it is important to become acquainted with the literature in range science and with methods of governmental agencies involved in measurement of rangeland vegetation. If there is some disagreement about the appropriate method, consider who will be the user of the final product, and let that dictate the field methodology. Measurements must meet three criteria of acceptability:

1. The method of making the measurements must be defensible.
2. The method must provide data that is representative of the population being measured.
3. The data must be in a form that is meaningful to the user of the final product.

MEASUREMENT ERROR

Although field personnel may never be required to provide a statement of the degree of uncertainty in their measurements, they should always be aware that error is likely to be present. Any given set of data may not be replicable by another operator, or even by the same operator using the same instruments. In addition to errors by the operator, other factors may contribute to the inability to reproduce data. Random elements may always be expected, and no amount of care can prevent them. For example, the exact starting location and direction of a transect is random, and the exact spot for collecting data along a transect may also be random. Only by collecting a sufficient amount of data can field personnel expect to compensate for random variation.

Another major source of uncertainty in measurements is operator bias. Theoretically, bias can be eliminated by setting guidelines for each step of measurement and by thorough training of the field personnel. Bias may be introduced by inadvertent selection of a "typical" shrub for measurement or by placing a quadrat frame so as to include, or exclude, some feature.

Field personnel should always plan on some practice time in the field area, or in a similar area. They should repeat practice measurements to assess the amount of variation they might expect in the data and, if possible, compare data with those taken by another operator. If they find large variations in subsequent measurements, they should try to tighten up procedures.

NOTE TAKING IN THE FIELD

Measured data or qualitative observations made in the field must be recorded in some manner suitable for future reference. Notes should be clear enough to be understood by anyone on the project—not just the field person. In addition to recording the data or observation of the features being mapped, information on slope angle and aspect and anything that may influence the spectral response at the site should be recorded. To facilitate note taking, it is useful to prepare a form for the purpose. Some examples are given in Appendix 2. These examples should be adapted to suit the needs of specific projects.

A useful component of field forms is the inclusion of the map category definitions directly on the form. This is a great aid to field personnel, and is especially important for maintaining consistency if there are several persons involved in making category decisions in the field. Also note sheets should identify any instruments by serial number. This may be useful information to other persons who may want to replicate or validate the data. Note taking requires considerable field time, and data forms should be designed to provide sufficient information without including unnecessary, time-consuming items.

FIELD PREPARATION CHECKLIST

It would be nearly impossible to provide an all-inclusive checklist of things to remember to take on every project field trip. Every project will have different requirements. However, certain categories of tasks, materials, and equipment can be identified. Project field personnel should always make a list of tasks to be completed before going to the field, as well as supplies and equipment to take to the field. A partial starter list for this kind of planning might be as follows:

Tasks

- Gather maps, imagery, and aerial photographs to take to the field.
- Make and field-check photo interpretation maps of cover type in the project area.
- Buy photographic film.
- Make certain equipment is operating properly.
- Make cluster map of the project area.
- Select control points for GPS.
- Select field sites from georeferenced imagery or cluster map, aerial photos, or field visits.
- Obtain permissions for trespass on private land. Failure to contact a landowner or resident could cause ejection from the land and block access to large parts of the study area.

Equipment and supplies

- Camera and film
- Equipment and instruments

  Tape
  GPS receiver
  Field spectrometer
  Compass
  Shovel
  Vegetation pruning clippers
  Secchi disk
  Munsell color charts

- Supplies and forms

  Field note forms
  Pens with several colors of ink
  Pencils
  Colored flagging
  Soil sample bags or cans
  Vegetation sample bags
  Water samplers
  Extra batteries for all instruments

- Personal items

  Appropriate clothing
  *Foot gear, and sun hat
  Sunscreen
  Sun glasses
  Drinking water containers (take plenty of water)
  Provision for meals and snacks

---

*Loosely fitted, long trousers and long-sleeved shirts protect the skin from sun, scratches from shrubs, and insect bites. In summer the skin actually stays cooler when protected from direct sun. Some field workers choose shorts for work in shrubby terrain. Avoid this sometimes painful mistake.

6

Measurement of Vegetation

SPECTRAL RESPONSE OF VEGETATION

Field personnel should have a good understanding of the spectral response of vegetation, including variations in response through the growing season, and with changes in moisture content. Knowledge of these variations aids in selection of the optimal time for field work and helps in selecting biophysical features for measurement. Figure 6.1 shows a generalized spectral response curve for vegetation from 400 to 2,500 nm. Note that the energy absorption to reflection relationships depicted in this diagram change with wavelength. Also, note that the biophysical controls (pigment, cell structure, and water) of the energy to plant interaction each affect different wavelengths. This can best be discussed by looking at three types of interactions of energy with plants—with leaf pigments, with leaf cell structure, and with plant moisture content (Swain & Davis, 1978).

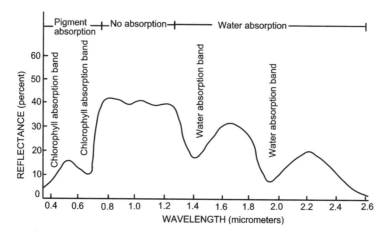

FIGURE 6.1. This typical spectral response curve for vegetation shows the characteristic bands that differentiate vegetation spectrally. From Hoffer and Johannsen (1968). Used with permission.

Leaf Pigments

Interaction of electromagnetic energy with leaf pigments is restricted to the visible wavelengths (400–700 nm). When chlorophyll is abundant in the leaf, it dominates both reflection (in the green) and absorption (in blue and red). This accounts for the two absorption bands on either side of the green reflectance band and explains why a leaf appears green.

When the leaf begins to change, due either to dehydration or senescence, chlorophyll production slows or stops, and reflection in the blue and red bands is increased while reflection in the green band is reduced. If other pigments remain in the leaf, they will dominate the reflection and the leaf will change color. Pigments such as carotene and xanthophyll reflect primarily as yellows and browns, and anthocyanin produces a strong red. In the visible wavelengths absorption and reflection at the leaf surface is dominant, and relatively little energy is transmitted through the leaves of most species. Figure 6.2 shows a generalized view of variation in spectral response of plants with change in moisture content. Note that reduced moisture in visible wavelengths results in an overall increase of reflectance. The loss of chlorophyll removes the absorption bands in the blue and red, creating a smoother curve in the visible. Loss of all pigments during senescence produces a high reflectance across the visible bands.

Information on the interaction of energy and plant pigments can help in planning field measurements. A field worker familiar with spectral curves of plant species would know that some species are almost indistinguishable by spectral analysis in the visible wavelengths. Another species may have less blue absorption and produce a slightly blue-green color, making it easily identified. Also, the low level of energy transmission through leaves in the visible wavelengths means that those wavelengths are not useful for making estimates involving plant volume, such as biomass or leaf area index. The visible wavelengths, however, serve well for estimates of plant coverage, such as percent cover or crown closure. Visible wavelengths will provide information related to moisture content only when a plant is under enough stress to stop producing chlorophyll and change color. By that time a plant may be near its wilting point.

Keep in mind that most information about plant spectral responses is based on studies of plants in humid environments. Less spectral information exists for plants in arid and semiarid environments. Also, even in humid environments considerable variation exists depending on season and moisture availability. When possible, field personnel would benefit from collecting field spectral information for themselves to determine variation among species and seasonal variation within species for a particular study area.

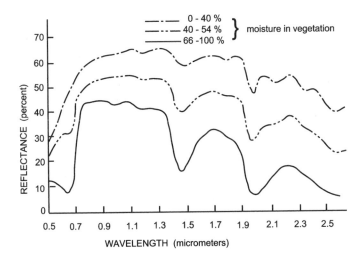

FIGURE 6.2. As vegetation dries, the percentage of moisture decreases; the reflectance becomes brighter and the water absorption bands become weaker. Absorption bands in the visible tend to disappear with drying. Adapted from Hoffer and Johannsen (1968). Used with permission.

Leaf Cell Structure

Beginning at 700 nm and continuing to 1,200 nm (see Figure 6.1), the nature of interaction of energy with plants changes significantly in several ways. The most obvious change is approximately a fourfold increase in the overall reflectance. This creates a very "bright" surface in the near infrared, and absorption becomes a negligible factor in the energy plant interaction. Another difference is that the interaction in the near infrared is determined by cell structure rather than pigments. A third difference is that transmission now plays a major role and accounts for nearly half the incident energy, equal to reflection, with absorption accounting for less than 5% of the energy. Only the first of these differences, "brightness," can be observed on the spectral response curve, but each of them provides a basis for what-to-measure decisions in the field.

The great increase in reflectance in the near infrared provides a remarkable capability for distinguishing vegetation from almost any other surface material, especially soil and water. This contrast is the basis for vegetation indices such as the tasseled cap transformation (Kauth & Thomas, 1976) and the perpendicular vegetation index (Richardson & Wiegand, 1977), each of which is based on the contrast of soil reflectance in the visible with reflectance of vegetation in the near infrared.

Not only is vegetation reflectance greater in the near infrared than in the visible, but the spectral response variation among plant species is also greater in the near infrared, especially among broad physiognomic categories (Figure 6.3). Greater discrimination of species in the near infrared allows field personnel to make more detailed observations about crops. Plants that have nearly the same response in the visible due to similar amounts of pigmentation show differences in the near infrared resulting from different cell structure in the leaves.

Spectral response also changes as cover density decreases and more soil shows through at the surface. Figure 6.4 shows that decreasing plant cover alters the spectral curve in stages from a typical vegetation curve to one increasingly like soil. This change in spectral response provides the basis for estimates of cover density from image data.

Energy Transmission through Leaves

Transmission of half the incident near infrared allows more energy in those wavelengths to reach leaves that are not exposed to an overhead view. Lower-

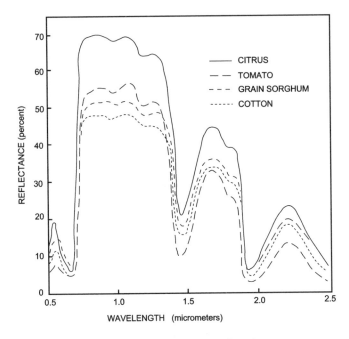

FIGURE 6.3. Reflectance from four different species shows strong separation of response curves in the infrared with weaker separation in the visible. From Myers (1970). Copyright 1970. Used with permission from the National Academy of Sciences, courtesy of the National Academies Press, Washington, DC.

level leaves then reflect half and transmit half of the energy they receive. The result is a useful accumulation of energy at the top of a plant much greater than the outer leaves alone would reflect. Thus, reflected energy in the near infrared relates to the mass of the plant—not just the crown diameter. Figure 6.5 illustrates the interaction that makes it possible to use near infrared wavelengths for measurement of biomass or leaf area index. This diagram shows that the total reflectance of a plant in the near infrared may be greater than only the reflectance of its outer layer of leaves. Each successive layer adds a decreasing increment to the total reflectance.

Plant Moisture Content

Figure 6.1 shows major absorption bands at 1,400 nm, 1,900 nm, and 2,700 nm. Absorption in these wavelengths is caused by water, and the amount of absorption varies directly with the amount of moisture in the plant. The value

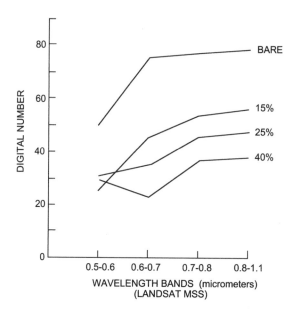

FIGURE 6.4. Variation in spectral response with vegetation cover decreasing from 40% to bare ground on a highly reflective desert soil. Typically soil would not be more reflective than vegetation in the near infrared. Note the response at 25% still shows a trace of the character of a vegetation response. The difference between 15% and bare ground is a matter of increasing brightness with a decrease in vegetation. The data are for four Landsat MSS bands and are not continuous spectra.

of this relationship is shown in the selection of wavelengths for TM with band 5 sensitive to 1,550–1,750 nm just on the overtone slope of a moisture absorption band. As a plant loses moisture, the overall reflectance increases across all bands, and the water absorption bands become less prominent (Figure 6.2). This feature provides a means of relating image values to moisture measurements of plants and soils. Field personnel could collect moisture samples from leaves, soils at the surface, and from soils in the root zone to compare with image data in the water absorption bands for the site.

WHICH BIOPHYSICAL FEATURES TO MEASURE

Plants provide a wide choice of characteristics that could be measured. Field personnel must decide (1) which plant characteristics affect spectral response in the available wavelengths, (2) which of those characteristics to measure to avoid redundancy and minimize field time, and (3) which measurement technique to use. Criteria for making these decisions depend on the intended final product. If the final product is a map, a preliminary legend for the map will have been devised in the statement of objectives, and this will control the selection of vegetation characteristics to measure. A useful approach for grasping this issue is to move from the most generalized observations to more complex measurements. This corresponds to requiring less field time or more field time to complete the work.

General Features

Even within a generalized level of information, steps of increasing specialization can be made. The most generalized observations require no actual measurements of vegetation. Simply identifying the physiognomic type (i.e., coniferous forest or herbaceous) should suffice. The next step in specializa-

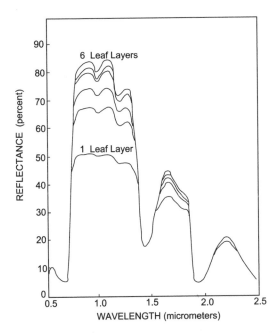

FIGURE 6.5. Reflectance from one to six layers of cotton leaves increases with each additional layer. This phenomenon is the basis for spectral estimates of biomass. From Myers (1970). Copyright 1970. Used with permission from the National Academy of Sciences, courtesy of the National Academies Press, Washington, DC.

tion, still requiring only observations, would be to identify dominant species in a physiognomic type with estimates of relative ground cover density (high or low) and proportions of physiognomic mixtures (grass vs. shrub). Much of this level of information may be derived from aerial photographs, but it should not be accepted without adequate field checks. Templates are available in aerial photography interpretation books and are useful for making estimates of tree crown closure when using aerial photographs.

If information apart from these visual observations and estimates is needed, then actual field measurements must be made. When project objectives require a more specific quantitative statement of cover density, with separation into cover classes, for example, then a technique involving transects or quadrats for measuring must be selected. A number of methods appear in the literature, and some of them will be discussed later. While collecting data for cover density, field personnel can add information on proportions of species within a plant community with only a moderate addition of effort and time. Methods for doing this will be discussed in the section on how to measure samples in the field.

Detailed Field Data

Only if the project objectives specify a need for a greater level of detail should field personnel consider the time and expense of acquiring data such as crown diameter and height, plant and soil moisture content, chemical analysis, leaf area index, or biomass. Each of these variables affects spectral response and can add valuable information to the final product, but each requires much more time to collect and analyze. The result may lengthen field time by many days. Avoid spending valuable field time collecting data that are not needed for a particular project. However, the anticipation of a possible future need for the extra data may justify collecting the data sooner rather than later, which would be too late for features as changeable as vegetation. Certain data collection, such as for moisture and chemical analyses, requires appropriate precautions to prevent moisture loss or chemical contamination of samples while transporting them to a lab for analysis.

TIMING OF MEASUREMENTS

In many areas of natural vegetation, as well as agricultural areas, seasonal timing of field work and image acquisition is important to the accuracy of the final product. Often remote sensing personnel select the "peak of green" period of the growing season. This time may give the strongest vegetative response and provide good data for biomass maximum; however it may not be the best time for discrimination of vegetation species by remote sensing. The reason is that vegetation species, whether natural or crops, often tend to have different growth characteristics such as early or later maturation across the growing season and show the least difference from each other at the peak of growth. This situation varies for each ecosystem and agricultural region.

The changes in spectral response result from phenologic changes through the season, beginning with sprouting, flowering, leaf development, and senescence. Each species in a particular study area may have a different annual schedule and reach its peak at different times in the growing season. Then all species may have a long period in the summer when all are at their peak, followed by varied times of onsets of senescence and harvest—in the case of crops.

Crop Calendar

A look at images for a particular study area will reveal if plant discrimination is difficult at certain stages of growth. Perhaps only two or three species cannot be separated spectrally. If such difficulty exists, then learning the phenologic details for a particular study area may be required. Making a phenologic calendar for natural vegetation, or a crop calendar for agricultural areas, is one way to approach this problem. Data for these calendars can be obtained from a literature search for ecological studies in the area in question and from interviews with experienced field-oriented ecologists in universities or in state or federal agencies involved with natural resource management in the region. Crop phenology for a given area can be learned from agricultural agents or from personnel at agricultural colleges or experiment stations. In addition, the U.S. Department of Agriculture (USDA) National Agriculture Statistics Service publishes information on crop conditions within each agricultural state. The object is to find the time when the look-alike plants are most different from one another. Figure 6.6 shows an example of a crop calendar for a hypothetical region.

Remote sensing workers may also apply classification techniques to improve accuracy where two or more plants cannot be reliably separated spectrally. Options such as contextual classification, a priori probabilities, or layered data may be used to improve classification results. For example, if the sites of one of the species have different physical characteristics, such as

	JAN	FEB	MAR	APR	MAY	JUN	JUL	AUG	SEP	OCT	NOV	DEC
CORN	stubble			planting		Increasing ground cover		tasseling	maturing	harvest	stubble	
PASTURE	dormant grasses				full cover green color		stays green in wet year turning brown in dry year				dormant grasses	
SOYBEANS			stubble			planting	Increasing ground cover			harvest	stubble	

FIGURE 6.6. Hypothetical crop calendar for an area in the midwestern United States. Note that the time in which corn fields and soybean fields are most different phenologically is in October during harvest. If those crops proved to be spectrally too similar to distinguish, then October imagery would provide a means of discrimination.

growing only on north-facing slopes or on a specific soil type, digitized slope or soil maps may combine as an additional data set along with image data to improve classification results. Refer to Jensen (1996) for details on these classification options. Experiments with classification options before field work begins should determine whether a crop calendar is needed.

Synchronicity of Field and Reference Data

The more dynamic the phenomena being observed or measured in the field, the closer in time the image data must be with the field work. Again, project objectives determine the answer to timing of field work with image date. If only species identification is the objective, timing of work and images should to be in the same growing season, particularly in agriculture, where crops rotate annually. If plant dimensions or volume, such as biomass, are needed, then be aware that early in the growing season plants change daily and a difference in field time versus image date of 1 week may be too much. Later in the season, plants are in full foliage, the rate of change slows, and field work time may differ from image date by as much as 2 weeks. Near the end of the season, change to senescence is again rapid enough that a few days makes a significant difference. Radiometric data cannot be entirely coincident with the time of overflight, but data collection should bracket the satellite overpass time so that the solar angle for the two data sets are approximately the same.

If field work involves moisture measurements in the plant or in the soil, imagery date and field work timing should coincide within 1 day. In arid climates a few hours may be significant for moisture change in the summer.

STANDARDIZATION OF MAPPING METHODS

Over the years many different methods for measuring and analyzing vegetation have been devised. Many workers are now concerned that too many methods exist. A reason for this great variety is that many possible objectives exist for any vegetation survey. The ecologist may have objectives different from a range specialist interested in grazing production. Both may have objectives that differ from those of a timber specialist. Furthermore, different plant communities may require different methods. One of the main factors preventing standardized methods from developing sooner is the constant effort by field scientists to find new techniques that make the most efficient use of field time, yet yield useful, accurate data. These efforts produce some new techniques that attract wide notice and many others that are seldom used except by their inventors. Most of the widely accepted methods for mapping vegetation were developed 25–50 years ago, and many of the more recent methods are modifications of earlier methods upgraded for greater speed or accuracy.

A movement is now under way to standardize vegetation analysis. U.S. government agencies concerned with vegetation mapping have begun to work toward standardization of methods. The federal government has initiated an effort to establish mapping standards for vegetation under the supervision of the multiagency Federal Geographic Data Committee (FGDC). The purpose is to create standards for all geospatial data to facilitate data interchange among agencies. The time when each agency has its own mapping standard, which may not be compatible with other agencies, is about ended. The FGDC has completed a draft document titled "Existing Vegetation Classification and Mapping Technical Guide," which can be seen on the U. S. Forest Service website at *http://www.fs.fed.us/emc/rig/*. Among topics covered in the FGDC draft guidelines are sections on vegetation map development and field data collection standards. The effort to standardize maps is commendable in these times of large, transferable geospatial databases. Also, the interchangeable databases make the development of metadata all the more important and useful to remote sensing personnel. Metadata resources are provided in Appendix 3.

Extensive field method instructions pertaining to forests, specifically forests of the western United States, are included in the document found in the

Forest Service website mentioned above. This lengthy document attempts to standardize plot measurement procedures, methods for measuring tree diameters and heights, and species names and vegetation classification. Anyone planning a forest vegetation map should examine this document.

GENERAL METHODS OF MEASURING VEGETATION

Vegetation field methods are based on two basic approaches: the line transect and the quadrat. Quadrats are often located along a transect line. This section will outline the basic techniques.

Line Intercept Method

When the field area is readily traversed on foot with no difficult topography or obstructions, the line intercept method of vegetation analysis is feasible. It applies to forests as well as rangelands. The method involves stretching a tape, graduated in feet or meters, in a straight line in a random direction from a random starting point. In order to keep the tape straight and taut during the survey, field personnel should stake each end securely.

The length of the tape depends on the density of vegetation and the floral diversity of the community. In a community of high cover density and relative few species (e.g., low herbaceous vegetation), the transects can be on the order of 10 m, with observations made every 10 centimeters (cm). In shrubs and trees, the length of the transect should be more than 100 m with observations every 1 m. In either of these examples 100 observations are made in each transect. If the community is layered, a transect for each layer is advised, and the resulting cover density for a particular area may exceed 100% (Daubenmire, 1959). As a practical matter, it should be possible to record multiple layers in a single transect by noting cover along a vertical line above the observation point.

For remote sensing purposes, observation points that fall under the canopy of a plant should be counted as an observation of that plant even though the plant's stem is not itself on an observation point. Unless a complete plant community composition analysis is needed, vegetation under the canopy may be ignored. Keep in mind that remote sensing field work is intended to evaluate what the satellite or aircraft sensor recorded. Cover density estimates are concerned with ground area covered from view from above.

The line intercept approach is not uniformly accurate in all types of vegetation. Best results are obtained in areas covered by shrubs and herbaceous growth. Accuracy improves with additional transects per site. The more diverse the flora or the sparser the cover, the more numerous the transects should be for a given sample site in order to assure representativeness of the results.

No rule of thumb exists for the right number of transects. Field personnel can see if each transect is showing additional species, or the same species in different densities. Additional transects may be made until the results appear to stabilize. If the overall cover of each species stays approximately the same with subsequent transects, then personnel may assume that the community is well represented and enough transects were made.

At each observation point the field person tallies the cover material at that point. Symbols for each cover type should be selected in advance. The letter *B* could mean bare soil, *L* for litter, *R* for rock, for example. When vegetation is at the observation point, the first two letters of the genus and species may be recorded (e.g., Artr. for *Artemesia tridentata*). If species are not needed, then the observer would record an appropriate symbol or term for shrub, grass, forb, tree. A field form for recording each observation in a transect should be set up in advance and tested in a field situation. Examples of field forms are provided in Appendix 2, but each project will require some adaptation of the forms to specific needs.

If 100 observations have been made on a transect, then the total of each cover type is also the percent cover for that type. For example, if a transect totaled 23 observations of bare soil, then the cover density of bare soil for that transect is 23%. The results of multiple transects for a single sample site (e.g., 3×3 pixels) can be totaled and averaged. The result provides an overall plant cover density, and a cover density for individual species. An estimate of species frequency may be computed as well.

A common pattern for sampling consists of a baseline of random direction staked at each end for identification. Transects then begin at the baseline, and at right angles to it. The length of the baseline depends on the size of the sample site, and a baseline may be established for each sample site. The distance between transects along the baseline is random.

One transect method frequently used by federal agencies in the United States is the Parker method (Stohlgren, Bull, & Otsuki, 1998). The Parker method (Parker, 1951) uses lines 100 feet long (30.5 m). Observations of cover type (species or bare soil) are made through 0.08-inch (1.9 cm) rings spaced 1 foot (30.5 cm) apart along a transect tape providing 100 points per transect.

The Pace Method

A variation on the line intercept method is the pace, or step-point, method. This technique applies the same concept without a tape. Observation points are determined by a mark on the toe-tip of a shoe sole as the field person paces along a transect. Some operators cut a notch in the tip of one boot. The observer records the cover at the point where the notch touches the ground. This approach may be applied in any vegetation, but is most suited for areas of shrubs and grasses of low cover density. The Bureau of Land Management uses this method for vegetation reconnaissance surveys covering vast tracts of semiarid and arid land in the western United States. The advantage is that large samples can be collected in a relatively short time. The disadvantage is that operator bias is present in each observation. The operator may unintentionally stretch or shorten a given step to select a plant rather than a bare ground observation. Although practitioners try to avoid this problem, the potential is always there and weakens the reliability of the survey. Nevertheless, the pace method stands as a suitable method for rapid vegetation surveys of large areas.

Another difficulty with the pace method is that the operator must occasionally detour around an obstructing tree or large shrub. Field personnel then create an offset around the obstacle for a portion of the transect line. The offset may introduce another source of bias. Multiple transects may help compensate for operator bias.

Field personnel in the Bureau of Land Management count one pace every time the left (or right) foot hits the ground, and observe the ground cover at the mark on the boot. They continue a transect until they have 100 hits. This results in a transect about 600-feet (182 m) long.

Recording observations and reducing results for the pace method are done in the same manner as described previously for the line intercept method.

Quadrats

Many of the sampling techniques based on observing quadrats utilize transects for locating the plots. One of those, widely used by federal agencies in the United States, is the Daubenmire transects method (Stohlgren et al., 1998). Though this method is called a transect, it differs from the line intercept method by observing small areas rather than points along the transect (Daubenmire, 1959). The Daubenmire procedure was later modified for U.S.

Forest Service use (U.S. Department of Agriculture, Forest Service, 1966). The method consists of transects 100 feet (30.5 m) long with 20 cm × 50 cm (0.1 m²) quadrats oriented perpendicular to the traverse at intervals of 5 feet (1.67 m). This provides 20 quadrats per transect. Daubenmire suggested that two such transects are adequate for many vegetation types.

Several tallies of plants within the quadrats may be made. Daubenmire identified canopy coverage as the most important, and included also the number of individuals by species, and dry weight (for computing biomass). Plants overhanging into the plot frame are included in canopy cover estimates but not included for counts of individual plants. For plant counts, the stem must actually be in the plot frame. Record keeping for this method is similar to the line intercept method, but needs to be adapted to accommodate additional information at each observation site, which are areas rather than points. The total area of quadrats along each transect is 2 m². in the Daubenmire method. Therefore, the sums of each variable recorded can easily be converted to amounts per square meter and projected to larger units such as hectares or acres.

Daubenmire suggested that the small quadrats make it easy to visually estimate the canopy coverage for each species into six classes based on percentage cover: 0–5, 5–25, 25–50, 50–75, 75–95, 95–100. He painted marks on the 20 cm x 50 cm plot frame to divide the frame into quarters. Then in one corner he indicated two sides of a 7 cm × 7 cm square. The marks then provide references for areas that are 5%, 25%, 50%, and 75% of the plot frame. The field person visualizes groups of plants fitting into the smallest reference area of the plot frame. For example, if one species in the quadrat could be visualized to fit into the 7 cm × 7 cm reference area of the frame, then the canopy cover estimate for that species would be 5%. Poreda (1992), following guidance of Professors Davis and Harper at Brigham Young University, modified Daubenmire's frame to a 1-m² plot frame, and added a reference area equal to 1% to improve canopy cover accuracy for infrequent species in an area of sparse vegetation (Figure 6.7).

Stohlgren et al. (1998) compared four rangeland vegetation sampling techniques to see how well they surveyed plant characteristics. The comparison included the Parker method, the Daubenmire method, the Modified-Whittaker multiscale quadrat method, and the "large quadrat" method. They found that the Modified-Whittaker (Stohlgren, Falkner, & Schell, 1995) method provided better results. The other methods underestimated the number of plant species. They attributed the difference to deficiencies in the transect methods. Those methods sampled too small an area and had a high

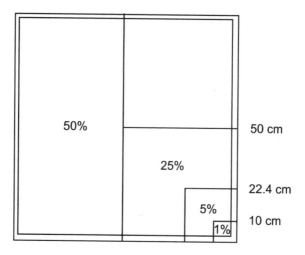

FIGURE 6.7. Plot frame with marks on sides to delimit segments of the quadrat into areas 1%, 5%, and 25% of the total area. Larger areas of 50%, 75%, and 95% can be estimated as well. From Poreda (1992). Used with permission.

degree of spatial autocorrelation due to the linearity and close spacing of quadrats. The Modified-Whittaker method provides nested plots with a large plot 20 m × 50 m enclosing one subplot of 5 m × 20 m, two subplots 2 m × 5 m, and ten subplots 0.5 m × 2 m.

Volume/Weight Analysis

It is common for remote sensing projects to require information on the volume of vegetative material present by growth type (physiognomy). This would be expressed as measures of biomass or forage weight for shrubs or grasses. Any of the quadrat methods described above can be utilized for dry weight measurements. Researchers have devised methods for estimating weight in a nondestructive way. Nondestructive, in this context, means that no harm is done to plants in the quadrat plot itself. Comparable plants outside the plot must be cut, dried, and weighed. The technique uses a reference unit cut from a nearby representative plant, such as a shrub. The method was described by Australian range scientists Andrew, Noble, and Lange (1979), and is called the "reference unit," or Adelaide, technique. Later Andrew, Noble, Lange, and Johnson (1981) compared the reference unit method with two other methods, volumetric measurements and capacitance probes. They concluded that the reference unit was more accurate than volumetric measurements and much less difficult than the capacitance probe.

Field personnel using the reference unit method to estimate biomass must select and cut a representative shrub branch from outside the quadrats. Operators move around each shrub making a visual estimate of the number of times the leaf and twig growth of the reference unit could be replicated within each of the shrubs rooted in the plot. (Andrew et al., 1979, were using 100-m^2 plots.) The same estimates are made also on several test plants outside the plots within a few hours, before the reference unit wilts. At the end of the survey period, the reference unit is stripped of foliage and new, current year twigs, and is then oven dried and weighed. The test plants are also cut, stripped, dried, and weighed. Biomass calibration between test plants and the reference unit is then applied to reference branch replication counts on shrubs rooted in the plots.

Gobena (1984) measured crown heights and diameters on shrubs in 100-m^2 plots to estimate biomass from crown volume. This method also requires cutting, drying, and weighing representative shrubs outside the plots to calibrate crown volume and biomass. He measured long and short diameters of the shrub crowns to compute an average diameter. The average diameter was also used to compute cover area of the crown. The sum of crown areas, by species, was the cover density of shrubs in the plot.

Another frequently used vegetation measurement is leaf area index. The computation of LAI is a simple ratio of total leaf area (single side of leaf) to the area of ground covered by the plant. As foliage increases, LAI also increases. Hence, LAI is related to measures of weight such as biomass and is used in projects concerned with questions of evaporation, transpiration, light absorption, or chemical element cycling. In herbaceous communities LAI can be determined by clipping vegetation from sample areas and spreading it on a measured area to arrive at a ratio of square meters of leaves per square meter of surface. Curran (1983) devised a technique of photographing the spread of leaves and projecting the picture on a wall with an array of random dots. The number of leaves contacting 100 dots was used to calculate the proportion of green leaf coverage per unit area of ground.

Computation of leaf area index in forests is more complicated. Typically, the approach is to measure the amount of light penetrating the forest canopy. Jordan (1969) describes a technique to determine the ratio of red wavelengths to infrared reaching the forest floor. The rationale is that the intensity of red light at the top of the canopy is slightly greater than the intensity of the near infrared. On the forest floor, however, the relative intensity of the infrared is many times greater due to the absorption of red by the leaves and transmission of infrared through the leaves. The greater the number of leaves, the greater will be the difference in red-and-infrared on the forest floor. Leaf area index can be measured throughout the forest after calibrating red to infrared ratios with direct measurements of LAI at several points. Measurements should be made only in direct sunlight when the solar angle is highest, for example, between 11:00 A.M. and 1:00 P.M. Manufacturers are now making photoelectric instruments designed for making such measurements.

Large amounts of data for leaf area index and fraction of photosynthetically active radiation absorbed by vegetation (FPAR) are now being collected globally and made available for users through agencies such as NASA and the Oak Ridge National Laboratory. Additional sources are included in Appendix 1.

SELECTING A MEASUREMENT METHOD

Selection of an appropriate vegetation survey method for a remote sensing project clearly requires some serious consideration. Options are numerous. They differ in detail and time required, and accuracy varies. Remote sensing personnel must select from these options a method that meets the criteria and objectives of the project. They must consider level of detail needed for the project, time available for field work, and expertise of field personnel. The options range from fast and coarse in the pace method to detailed and time-intensive in the multiscale, nested quadrat approach. The fastest method, the pace method, is sufficient to identify major species and make estimates of broad classes of cover density of vegetation versus bare soil in a short time. If more information on other species is needed, the line intercept using stretched tape may be more reliable without too much additional time. Projects needing more detail require selecting a quadrat method suitable to time constraints and accuracy expectations.

7

Soil and Other Surface Materials

SPECTRAL RESPONSES OF EARTH MATERIALS

The spectral response curves of soil, rocks, and minerals are not so distinctive as those of vegetation and water. They tend to produce a rising curve through the visible and near infrared, and may continue rising, but less steeply, from wavelengths 1,000 through 2,500 nm (Swain & Davis, 1978). At first appearance, the only distinctive characteristic may be the water absorption bands at 1,400 and 1,900 nm. Even water absorption bands tend to diminish as a soil dries. The lack of complexity in soil spectral responses is the feature that makes soil and rock so easily distinguishable from vegetation or water. The various vegetation indices are based on this contrast of spectral responses for vegetation and soil. Stoner and Baumgardner (1981) identified five distinct spectral response curves for soils, depending on relative concentrations of organic matter and iron content along with variations in texture.

The characteristic pattern present on spectral plots of near infrared data plotted against red band data shows vegetation points with low values in the red and high values in the infrared (Figure 7.1). The so-called soil brightness

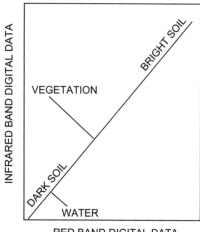

FIGURE 7.1. A plot of red band and infrared band digital data gives a soil brightness line with vegetation points falling on one side of the line. Used with permission of the American Society of Photogrammetry and Remote Sensing, A. J. Richardson and C. L. Wiegand, Distinguishing vegetation from soil background, *Photogrammetric Engineering and Remote Sensing, 43,* 1547–1552.

line, as described by Richardson and Wiegand (1977), separates vegetation points from water points on the plot. The soil line may include both dark and light soils, and can include asphalt and concrete at the extreme dark and bright ends, with gradations of soil or rock brightness in between. For many remote sensing projects it is enough to identify a material on or near the soil line by some name from the map legend, for example, asphalt road or bare soil. Other projects require more detailed information on soil, and for that purpose project personnel need to understand more about the components of the soils that contribute to their spectral response. At the very least one should understand the effects of moisture, organic matter, soil texture, or iron oxide content. Although the general appearance of most soil reflectance curves appears similar, these properties of the soil contribute significantly to response curves. The most apparent of these is soil moisture.

Visible Wavelengths

Soil Moisture

Moisture in a soil is always apparent by the pronounced absorption bands at 1,400 nm, 1,900 nm, and 2,700 nm. Two effects appear as a soil dries.

Reflectance increases and the absorption bands become less pronounced until they essentially disappear in the spectra of sandy soils. As clay and silt content increase, the moisture absorption bands diminish, but remain visible during drying. The progression of this change is regular enough to show a clear relationship between spectral response data and moisture content across the response curve, but particularly in wavelengths near the water absorption bands (Hoffer & Johannsen, 1968; Bowers & Hanks, 1965). Figure 7.2 illustrates the effect of drying on the spectral response of sandy and clay soils. Note that the driest clay soil still shows evidence of water absorption, while the dry sandy soil shows no trace of water absorption.

Visual experience tells us that moisture also reduces reflectance in soils in the visible wavelengths, and soil appears darker. This applies to moisture

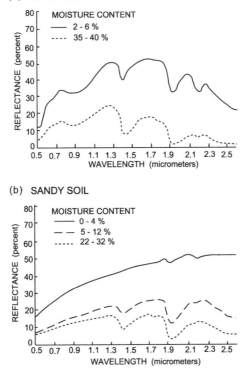

FIGURE 7.2. Spectral response of soils changes with the loss of moisture. Note that dried sandy soil (b) shows almost no indication of the water absorption bands, while dried clay soil (a) still shows some effect of the water absorption. From Hoffer and Johannsen (1968). Used with permisision.

on rock, concrete, or asphalt as well. A difference in the visible is that once soils are moistened, reflectance does not continue to decrease as more moisture is added, so the relationship between moisture and reflectance is less sensitive to the quantity of moisture, except to show the contrast between dry and wet.

Soil Texture

Particle size is a major factor in water retention in soils, so the strong relationship between spectral response and moisture is partly the effect of texture. This relationship is apparent when clay soils dry, but the effect of water on the absorption bands is still present. The larger particles of sandy soil, on the other hand, retain little moisture, and the absorption bands may disappear.

Texture, however, has an influence on spectral reflectance independent of moisture. If other factors are held constant, reflectance tends to be higher as particle size decreases and the surface becomes smoother. Bowers and Hanks (1965) showed that sand at 2.65 mm absorbed 14% more radiation than silt at 0.022 mm.

Organic Matter

Soil color is greatly affected by small amounts of organic matter. A soil may appear black with as little as 5% organic matter content (Swain & Davis, 1978). As organic matter decreases, soil color lightens to shades of brown or gray. The effect of organic matter on spectral reflectance is apparent only in the visible wavelengths.

Reflectance decreases with increasing organic matter. The limit to this relationship occurs when organic matter reaches about 5%. Above that concentration no further variation in spectral response due to organic matter is detectable. Climatic regions and drainage conditions each have significant influence on soil color and further complicate the relation between spectral response and organic matter (Stoner & Baumgardner, 1981). These factors must be taken into consideration on a case-by-case basis.

Iron Oxide

The visible wavelengths respond to changes in iron in the soil. Aside from the color iron oxides add to soil, other effects of iron can be detected with hyperspectral data. Broad absorption bands occur for ferric iron and ferrous iron centered at 560 nm and 830 nm respectively (Hunt, 1977). These bands can be used to detect anomalous concentrations of ferrous or reduced iron. Sites where reducing environments exist may result from exclusion of oxygen due to poor drainage, or from chemical reduction due to a subsurface presence of sulfide ore bodies or petroleum. These anomalies can be detected by ratioing the appropriate absorption band against a standard wavelength (e.g., 1,650 nm) that is unaffected by most mineral absorption bands.

Reflective Infrared (1,000–2,500 nm)

Clay Content

At wavelengths around 2,210 nm clay minerals are absorptive of radiation. Several distinctive characteristics in that wavelength zone make it possible to distinguish certain specific clay minerals (Hunt, 1979). As with the iron, ratios are a common practice for detecting areas of clay alteration related to anomalous sites due to hydrothermal alteration or reducing environments.

Calcite

A distinctive absorption band due to calcite occurs at 2,330–2,350 nm. In arid regions calcite may exist in the soil in many locations, and by utilizing its absorption characteristics one may identify variations in relative concentration.

Hydrocarbons

Hydrocarbon gases adsorbed to clay particles in the soil can be detected using spectrometers with hyperspectral capability. The response curves of methane and crude oil are very similar, with absorption bands at 1,180 nm, 1,380 nm, 1,723 nm, and a broad band at 2,300–2,450 nm (McCoy, Blake, & Andrews, 2001). Figure 7.3 shows that the two bands at 1,180 and 1,380 nm are sharply defined but fairly weak, with about 20% absorption. The 1,380 nm band is often obscured by water absorption at 1,400 nm. The band at 1,723 nm is stronger at 65% absorption, and the strongest absorption is the 2,300–2,450 nm band, with up to 95% absorption. Separately these hydrocarbon absorption bands overlap with certain mineral absorption bands. Most notably, calcite overlaps partially with the 2,300–2,450 nm band, but the calcite band is much narrower. Taken together, the hydrocarbon absorption bands form a definitive spectral indication for the presence of hydrocarbon gases in the soil. The

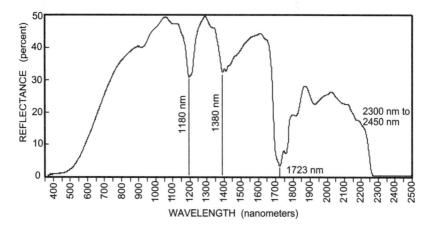

FIGURE 7.3. Spectrum of crude oil, showing absorption bands characteristic of hydrocarbon gasses adsorbed to soil particles.

potential use of hydrocarbon absorption in soils for environmental or exploration uses merits further development.

Other Minerals

Many minerals occur in soil and rocks visible at the surface. With the vast improvements of hyperspectral sensing, both in the field and by satellite, minerals can be sensed if they have distinctive absorption characteristics in their spectral responses. A large collection of mineral spectra created by the Spectroscopy Laboratory can be viewed at the USGS Spectroscopy Laboratory website, *ttp//speclab.cr.usgs.gov/*.

WHAT SOIL FEATURES TO MEASURE

Project objectives remain the primary guide to the selection of features for field observation or measurement. Many projects concerned only with cover type may require no information about soils, except to identify bare ground not covered by any other material. The fact that a site lies somewhere along the soil line in a red/IR plot may be sufficient. If more soil information is needed, then remote sensing personnel must select the level of detail required to meet the objectives of the project.

The lowest level of detail may consist of simple estimates or observations of texture, organic matter, color, or other features of the surface soil only. As the need for detail increases, the field time required increases by big jumps. At a maximum level of detail, field personnel will be digging soil pits and analyzing texture, structure, and soil chemistry for each horizon from the surface down to parent material, similar to the county soil surveys done by the USDA Natural Resources Conservation Service (formerly Soil Conservation Service).

Lowest Level of Soil Detail

Assuming that some analysis other than noting the fact that bare soil is present at the surface is needed, remote sensing personnel may wish to investigate why one soil is more reflective than another. This seemingly simple question opens the door to collecting data on soil moisture (though only if field work happens to coincide with overflights), soil texture, organic content of surface soils, surface soil color, and slope and spectral response of surface material, if a spectrometer is available. Even with this amount of required information, one may opt either for quick estimates made in the field or more detailed and quantified data measured on field samples taken to the laboratory. Each option is decided on the basis of the importance of soil information to the overall objectives of the project.

FIELD ESTIMATES OF SOIL CHARACTERISTICS

Topography, drainage, and vegetation cover should be the first observation around the immediate area at a field site. The main topographic variables to note are slope angle and aspect. Both affect the reflectance of a surface as well as soil characteristics and moisture. Slope and aspect can be estimated visually into broad categories, or with a compass and an inclinometer (both contained in a Brunton compass) quantified data on slope and aspect may be measured quickly.

Soil Color

Soil color is a characteristic that can be made easily on site, and color has a strong relationship to variations in soil spectral response. Color determination is best done with a Munsell® (1973) color chart. The chart consists of 196 color chips and alphanumeric notations based on hue, value, and chroma. The field person places a soil sample behind the appropriate color chip so it shows through the opening adjacent to the most similar color chip (perfect color matches seldom happen). Then color notation is made by hue, written at the top outer corner of each page. Next the number for value is written. A value of 6 means the brightness is 60% of the range from black to white. Then chroma shows the amount of color added from gray to full color. Hence, 5YR 6/4 is a color description that indicates a soil about halfway along the color gradation from full yellow to full red (5YR). The value, 6, is 60% of the scale from black to white, and the chroma of 4 shows a color intensity on a range from 0 to 20. No soil actually reaches a chroma of 20. Rather, a maximum for most chroma is about 8. The soil in question might be called "medium brown," but such a term is far too ambiguous to have any real meaning.

Full instructions for use of the Munsell charts are included at the front of the Munsell book. Field personnel should practice using the color charts before going to the field. Field notes should include information on whether the soil was moist or dry at the time of color designation. If soils are moist in the field when colors are determined, samples should be taken to the lab for drying, and a second (dry) color designation made.

Soil Texture

Soil texture may be estimated by feeling soil with the fingers, sometimes called the "feel method." This method provides an approximation of texture that may be suitable for many projects. With a little practice field personnel can learn to categorize texture on two scales, grittiness and plasticity. Grittiness, ranging from very gritty to smooth, describes the amount of sand (gritty) and silt (smooth) size particles. Old field hands then give the soil a spit test for plasticity. They wet a small sample of soil in their hand, and attempt to roll it into a string by rubbing their palms together. The more clay in the soil, the longer and more cohesive the string will be. If no string forms (no plasticity), then clay is low to absent, and the soil would be called sand or silt depending on the grittiness only. If a soil feels gritty and forms a string, then it would be called a sandy clay. Smooth-feeling soil that forms a good cohesive string would be called silty clay. Refer to Figure 7.4 for other possibilities. Practice with sieved soils of known texture gives added confidence to soil texture tactile classifications made in the field.

Surface Roughness

Surface roughness in the form of stoniness, organic litter, or eroded rills on the surface also affects reflectance. Therefore, estimates of the extent of sur-

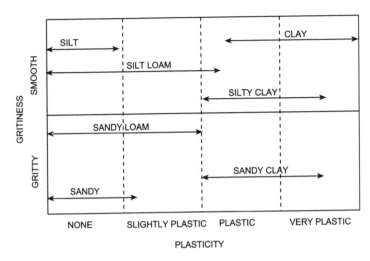

FIGURE 7.4. A soil texture classification made in the field is based on the feel of wet soil to the fingers and the ability of the particles to adhere in a rolled "string."

face cover by stones should be made, along with the approximate size of stones. Likewise litter, such as fallen leaves or twigs on the surface, influences reflectance, and its extent of cover should be observed and recorded when variations in surface reflectance are under study.

The extent of surface erosion may also be critical to surface reflectance, and should be noted at each site. Instances of severe surface erosion are evidenced by pebbles on pedestals cut into soil unprotected by vegetation from rain-splash erosion. Rills are evidence of unrestricted runoff erosion. Both features produce a surface roughness that is important to the study of surface reflectance. Notations should be made in terms of the extent of surface area affected by various contributors to surface roughness.

Tillage will also affect surface roughness. Plowing a moist soil produces a very rough surface, and subsequent harrowing will make the surface smooth, causing the same soil to have different spectral responses at different times.

Soil Moisture

Soil moisture can be estimated on-site with a simple resistivity moisture probe available at garden stores. Such probes are about 25–30 cm long, and can sense moisture at the surface or penetrate into the soil. This approach provides only relative moisture (wet, moist, dry, very dry), but may be adequate for some applications. The common practice for more precise moisture data requires collecting soil samples from appropriate depths (e.g., surface, root zone, or each horizon) and measuring the weight difference after drying as a percentage of the dry soil weight. The sample must be contained in a sealed metal sample can or in an airtight plastic bag until it is weighed before drying. Drying should be done in an oven at a controlled temperature of 105°C. A good reference for this and other soil analysis methods is Black (1965).

If repeated moisture measurements will be needed over a period of time, one should consider installing a device that stays in the soil and can be monitored periodically. Tensiometers, plaster, or nylon units buried in the soil come into equilibrium with the soil moisture over time, and water levels or resistivity readings can be made during field visits. These devices are set to a particular depth, and if moisture data at several depths are needed, then additional devices must be left in the soil at those depths. Field personnel must consider the trade-off of time spent taking soil samples at various depths on each field visit, versus installing semipermanent moisture sensing devices for the duration of the project.

Organic Matter

Soil organic matter may be estimated in the field, but the procedure is best in a laboratory since it involves the use of a strong base (sodium hydroxide). Bowman, Guenzi, and Savory (1991) provide details for a field technique that requires separation of the organic matter chemically and evaluation of the extract color density either visually or by spectrophotometer. Most remote sensing field personnel are not likely to be equipped or trained sufficiently for this procedure.

Soil color alone is not a reliable way to estimate organic matter content. However, in a small area where the type of organic matter is similar and mineral matter has a relatively uniform composition, color value may be used as a rough indication of organic matter content. In such a situation, assuming other factors to be constant, darker soils would indicate greater relative quantities of organic matter.

SOIL MAPPING

Much of the soil in agricultural lands of the world has been mapped by government agencies. More-remote areas may have some soils maps completed, but the coverage is spotty and priority for additional mapping is usually low. The technique for soil mapping requires personnel with expertise in soils to visit the field area to be mapped to become familiar with the area's soils and their relationship to topography, vegetation, and drainage patterns. They will then use interpretation of aerial photographs to delineate boundaries for soil types observed in the field. Boundaries between soil types are usually indistinct, and usually cover a transition zone. Most soil boundaries are determined by other physical features, such as vegetation changes, breaks in slope, or changes in drainage patterns that may be visible on the photographs. The next step is digging soil pits and describing the characteristics of each soil horizon from the surface down to the parent material.

If maps are made from satellite imagery rather than aerial photographs, mapping terrain units may be the best approach. Terrain units are spectrally similar areas that may be expected to have similar vegetation, topography, moisture conditions, and soils. The first step involves image interpretation to delineate terrain units on the basis of spectral regions. This procedure results in a visually produced cluster map. One may also try the computer approach to cluster analysis. As in any cluster analysis, field work is the next step.

Field checking terrain units (clusters) often will not reveal a uniformity of any physical aspect within the unit. There will be variation of soils, vegetation, and topography. However, when visiting subsequent field sites, one may see that there is less variation within terrain units than between adjacent terrain units. If that is the case, field personnel should move to the next step, which is to characterize the physical features of each terrain unit. This will require collecting data on soils at various places within each unit as well as gathering information on vegetation, moisture, and topography. Soil pits may be dug to obtain information through the horizons, and chemical, mechanical, and mineral analyses may be made.

Maps of terrain units are much less precise than aerial photography maps focused specifically on soils, and should be regarded as a reconnaissance tool. Such maps should not be called soils maps even though much soil information has been collected. Producers of terrain unit maps must be careful not to oversell their usefulness. Users must recognize that great variations may exist within each terrain unit, depending on which element of the physical environment is dominant in the spectral character at a given location. In most cases soil will not be a dominant influence on the spectral response of an area but will have an underlying influence on vegetation and moisture conditions. Hence, one aspect of the environment, such as soil, may vary without a corresponding change in vegetation.

LOCATIONS OF SOIL SAMPLE SITES

Field personnel, as always, need to be aware of project objectives when selecting locations for collecting soil samples or digging soil pits. Objectives may focus on vegetation, with soils or moisture considered as correlative elements only. In such a case, soil would be collected along vegetation transects or adjacent to quadrats. If the objective is to make a map of soils, then sites for several pits should be selected and dug within each soil mapping unit. Alternatively, transects of augered holes can provide much of the information obtained from pits in much less time. Be aware that augered soil data provide less accuracy on horizon thickness and no information on soil structure. A recommended procedure is to combine augering with soil pits on a ratio ranging from 2:1 up to 40:1, depending on the accuracy and density of soil data needed for the project (Breimer, van Kekem, & van Reuler, 1986).

If finding chemical or mineral anomalies is the objective, then field personnel should collect soil samples along a series of parallel traverses, with samples collected at intervals of a few meters up to a few hundred meters, depending on the anticipated size of an anomaly. This produces a grid that allows the resulting chemical or mineral concentrations to be mapped with isolines. Hyperspectral image data may then be employed to detect the same anomalies through analysis of appropriate absorption bands.

8

Water Bodies and Snow Cover

SPECTRAL RESPONSE OF WATER

The spectral response of a water body is a function of water characteristics, depth of the water, and organic or inorganic matter suspended in the water. Compared to soil or vegetation, the transmission component of the energy–matter interaction is much greater in wavelengths shorter than 760 nm. As a result much of the visible reflection component from water originates within the volume of the water body or from the bottom. Wavelengths longer than 800 nm are almost entirely absorbed by water regardless of its depth, chemical composition, or suspended matter, so reflection and transmission are not significant in those longer wavelengths (Swain & Davis, 1978).

Reflection within Water and at the Surface

Reflection from the surface and upper volume of clear water is spectrally similar to sunlight, except for reaching zero near 760 nm and beyond. The peak of the reflectance curve, like sunlight, is about 580 nm and reaches a maximum of about 6% of the incident radiation. Figure 8.1 shows a typical diffuse reflection response from clear water with little suspended sediment. Water also produces a specular reflection, especially as sun angle decreases, making a bright sun glint. As sun angle decreases, the proportion of transmission also decreases. Figure 8.1 also shows the effect of increasing turbidity. Suspended soil in the water causes the peak of the reflectance curve to shift into the red, and reach a reflectance near 10%. The response of turbid water takes on a response characteristic of the material in suspension, usually soil. Note also that when suspended material is present the near infrared reflects a bit farther into the near infrared wavelengths. This results from very shallow penetration of near infrared radiation, which is reflected from suspended solids near the surface. If the suspended matter in the water is algae, reflection will shift toward the green and peak near 580 nm, but it would be difficult to distinguish algae from suspended sediment if both are present because the stronger reflection and broad peak of the soil response curve in visible wavelengths would obscure the reflection from the algae. When floating on the water surface, algae reflects similarly to other vegetation.

Reflection from the Bottom

As radiation transmits into clear water, absorption and scatter attenuate the energy. Wavelengths longer than 900 nm are absorbed in the top few centimeters. Between 700 and 800 nm energy will penetrate as much as 1 m before being completely absorbed. Then the red wavelengths are absorbed within about 4 m. For mapping the bottom of a body of clear water, wavelengths between 450 and 550 nm provide the deepest penetration. The wavelength peak of energy penetration in clear water is 480 nm and can reflect from the bottom up to 20 m deep (Swain & Davis, 1978).

When the water contains suspended sediment, scatter and absorption increase, and much of the energy reflected to a sensor is from the sediment. In turbid water the reflectance peak shifts to longer wavelengths and peaks between 550 and 600 nm. As turbidity increases, the depth of energy penetration decreases, and the ability to map bottom reflection is reduced. Figure 8.2 shows the relationship of energy transmission to wavelength for clear and turbid water. At some point the reflection is entirely from suspended sediment, and none is from the bottom. In shallow water it is often difficult to know if the reflection is from the bottom or from suspended matter in the water. For high concentrations of sediment it is best to assume reflection is from the sediment.

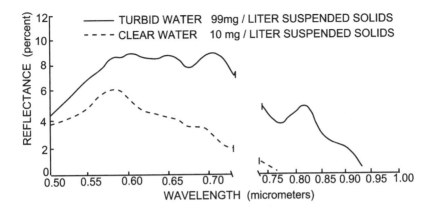

FIGURE 8.1. Spectral characteristics of turbid river water and clear lake water. Gap in data is due to instrument limitations. Used with permission of the American Society of Photogrammetry and Remote Sensing, L. A. Bartolucci, B. F. Robinson, and L. F. Silva. Field measurements of the spectral response of natural waters, *Photogrammetric Engineering and Remote Sensing, 43*, 595–598.

WHAT CAN BE DETECTED, MEASURED, AND MAPPED IN WATER?

Material floating on the water surface is usually easy to detect and to map its areal extent. Algae, for example, appears prominently in near infrared as a bright surface on a dark background. Floating oil is detectable, though not so prominently, in shorter wavelengths, 350–450 nm, and again in the thermal infrared.

Below the surface, suspended particles of sediment or algae may provide sufficient response, depending on their concentrations, to be detected and mapped. The bottom of the water body, up to 20 m in clear water, can be detected for mapping water depths. The most obvious feature, of course, is the areal extent of the water body that contrasts strongly with the land, providing a distinct shoreline. Certain things cannot be detected. This includes gasses (oxygen, carbon dioxide) dissolved in water, inorganic salts (sodium chloride), and acidity (pH).

TIMING OF WATER MEASUREMENTS

Water bodies are some of the most dynamic features on earth, and certain aspects of them may change in a day, or even a few hours. Lakes and reser-

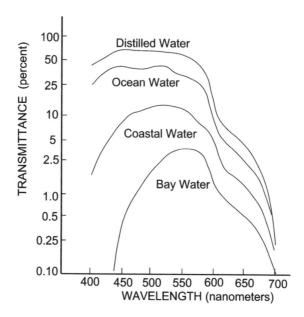

FIGURE 8.2. Spectral transmittance for 10 m in various water types. Used with permission of the American Society of Photogrammetry and Remote Sensing, M. R. Specht, R. D. Needler, and N. L. Fritz. New color film for water-photography penetration, *Photogrammetric Engineering and Remote Sensing, 39*, 359–369.

voirs may change level rapidly during high water runoff periods, or again during drawdown periods. The rate of change in water level will vary with the area and volume of the water body and the shape of the basin containing it. Flooding is usually a short-term event, and flood monitoring by remote sensing requires frequent images, though little field work.

Sediment concentrations—and algae—are not uniformly distributed in lakes and reservoirs, and significant areal variation in concentration is often present. Currents in a lake can be due to wind and stream inflow, producing a visible pattern in the suspended sediment that changes continuously. Even in rivers, where stronger currents and turbulence provide more mixing, sediment concentrations are not uniform either laterally or vertically.

All these variables make field work on water bodies highly time-dependent, and may require making observations and taking samples at the exact time of an imaging overflight.

FIELD MEASUREMENTS

Suspended Sediment and Turbidity

The main source of suspended sediment is erosion of soil and transport by runoff into streams and lakes. Much of this material may settle quickly to the bottom of the water body and become bed load. The rest is finer silts and clay particles that remain suspended in the water for long periods. Remote sensing projects concerned with water often attempt to monitor suspended sediment concentrations because it can be sensed by reflection of green and red visible energy from sediment suspended within the water. It is possible to distinguish imaged differences in sediment qualitatively without field work simply because higher concentrations of sediment create brighter responses along with wavelength shifts from 480 nm in clear water toward 550 nm in turbid water.

Sediment in the water may be observed in two ways: as suspended sediment measured in weight of sediment per unit volume of water, or as turbidity measured in terms of light transmission through the water. The terms "suspended sediment" and "turbidity" refer to the same phenomenon but are not interchangeable. Suspended sediment is the concentration of particles suspended in the water. Turbidity is an optical quality referring to the amount of light transmitted through the water. Normally a positive correlation exists between suspended sediment and turbidity, leading people to treat the terms interchangeably.

Suspended sediment may be collected in a sample bottle designed so that the operator can open and fill it, at a selected depth. Depth-integrating sample bottles fill gradually as they are lowered into the water and provide a single sample representing a range of depths. The water samples are taken to a laboratory for analysis. Also, it is possible to measure suspended sediment *in situ* using a sediment gauge. The sediment gauge operates within the water body by passing visible light of constant intensity through the water to a detector at a fixed distance from the source light and recording the light attenuation by suspended sediment particles. Since the sediment gauge is actually recording turbidity, a number of bottled samples must be taken simultaneously for calibration to convert the data to suspended sediment units, for example, milligrams/liter.

Turbidity is measured according to the attenuation of light transmitted through the water (absorbiometric), or it may be measured by the amount of scattering as light passes through water (nephelometric). Absorptiometers are used for samples of high turbidity, but in relatively clean water a nephelometer provides better sensitivity. Different results may be obtained from instruments of the various manufacturers, so turbidity data are treated as relative values and may not be comparable to data from other sources. In either case these are laboratory instruments for use on water samples brought in from the field.

The Secchi Disk

A useful device for evaluating water transparency is the Secchi disk. A 20-cm white, with alternate black quadrants, metal disk on a measuring tape or cord is lowered into the water with the flat side parallel to the water surface. A weight should be attached below the disk to ensure that the measurement is vertical. The operator makes note of the depth at which the disk disappears from view. The Secchi disk operator can check the depth reading by lowering the disk until it disappears in the turbid water, and then raise it slowly until it is just visible again, and average the two readings. Some users recommend a piece of 10-cm-diameter plastic pipe about 1 m long as a viewing tube. The measuring tape holding the disk need not pass through the tube. Viewing the Secchi disk through a tube inserted about 30 cm into the water allows the operator to see the disk without the glare of sun on the water surface. If no tube is available, take all observations on the shady side of the boat. However, even on the shady side, skylight causes some glare.

Using this disk has the advantages of simplicity and immediate results. The data are an indirect measure of water transparency and are not the same as direct measurements of suspended sediment or turbidity. However, unless those direct measurements are specifically needed, a remote sensing project can easily justify using the Secchi disk. The disk provides information on the depth of light penetration in a given water condition. As light penetration in the visible wavelengths is also the phenomenon recorded by remote sensors over water bodies, it is logical to use the Secchi device. If data for actual values of suspended sediment or turbidity are needed, some simultaneous water samples may be taken at the Secchi disk depth for calibration.

The Secchi disk itself provides no information about the reason for variation in water clarity. Suspended particles may be soil or algae, and samples should be taken to evaluate which are present. Water color is also a factor determining the depth of visibility of the Secchi disk, and recorded observations should include notes on water color that may relate to types of algae present.

When submerged in very clear and deep water, the disk may disappear due to insufficient light at depth. Changing to a larger, all-white, disk of 60 cm diameter or more can correct the problem, but in remote sensing work that should not be necessary. In a remote sensing project, one can assume that data from deep clear water is not recording more than 20 m depth. In the clear water of Crater Lake, Oregon, for example, a 20-cm disk disappeared from sight at 39 meters. At the same time a 1-m disk was visible down to 44 meters, which is the record for a Secchi disk in lakes.

Several operational procedures should be followed when using the Secchi disk. Davis-Colley, Vant, and Smith (1997) recommend allowing 2 minutes of viewing the submerged disk through a viewing tube, to enable the eyes to adapt to the light before recording the depth of disappearance and reappearance. Also, readings should only be taken during the high sun period when light penetration is greatest. Between 10 A.M. and 2 P.M. is the optimal period. The angle of sunlight entering the water is critical to the consistency of data using a Secchi disk. If sun angle (due to time or latitude) is not considered, the Secchi depth may appear shallower, and the operator may attribute different Secchi depths to suspended material in the water.

Verschuur (1997) demonstrated that Secchi depth varies with sun angle, and he devised a calculation for normalizing readings to the value that would be obtained if the sun were at the zenith. Without such normalization Secchi data cannot be compared over time, or from different lakes at a given time. For best accuracy and precision 6–10 Secchi measurements should be averaged and corrected for solar altitude before attempting to compare data from different lakes. The zenith correction is given by the equation

$$Zsd = Asd \ (1 + F) \ / \ 2F \qquad (1)$$

where Zsd is the Secchi depth corrected to zenith value; Asd is the apparent, or observed, Secchi depth; and F is a scaling factor determined by the solar altitude, a, and computed by equation 2.

$$F = \cos a \ [\sin a - 1 \ (\cos a \ / \ 1.33)] \qquad (2)$$

Sampling Plans

The spatial patterns of variation of sediment in a water body on the day of field work cannot be known by observation from the shore. This means that field personnel cannot see in advance which sites should be sampled. Therefore, a systematic plan for collecting data must be devised. It helps to have an image of some previous date to get an idea of the magnitude of the pattern of variation that might occur, keeping in mind that the pattern will vary somewhat on another day. An image from an earlier date will help in determining the magnitude of the patterns of variation, and provide a guide to the necessary frequency of observations. Several events may change the sediment conditions over a few days' time. Major storms in the watershed will bring new sediment into streams and lakes, and seasonal recreation may stir up existing sediment. Seasonal changes in algae growth can also change the pattern of suspended matter in the water. These elements should be kept in mind when looking at images from a previous date.

The sampling plan should consist of several transects across the water body collecting data or samples at regular intervals along the way. A GPS receiver is essential on a water body in order to know when one has reached a selected sample site. Plan on making enough observations with spacing close enough to have 6–10 points in various sediment concentrations. One point in each level of concentration might seem to be enough, but additional points provide greater confidence that the data are representative of sediment concentrations seen as reflectance variations on the image. Data points along the traverse should be no more than 1,000 m apart, and if spatial variation is likely to be high, 200 m should be considered. Transects may be 1,000–2,000 m apart. If data will be contoured, the closer spacing of transects will provide a better map. All these suggested sample plans may be adjusted, depending on the dimensions of the water body and the anticipated magnitude of the sediment pattern. If large areas need to be covered in a limited time, field personnel could make fewer transects, but the spacing between sample sites should not be expanded. If time is limited, strive for confidence in data in a smaller area rather than data of uncertain value over a large area.

Collecting Suspended Sediment Samples

Equipment for sampling suspended sediment consists of collectors of various designs having about 0.5 liter up to 1 liter capacity. The design provides sufficient weight to submerge the sampler and open the container at a desired depth, and to allow a smooth, nonturbulent flow into the sampler. A container inside the sampler is easily removed and taken to a laboratory for analysis. In the laboratory the containers are allowed to stand for 1 or 2 days. The water is then carefully drained off, the remaining sediment oven dried at about 110°C and weighed. One variation is to allow the water to evaporate rather than draining it. This has the advantage of including particles in colloidal suspension, but also includes dissolved solids. If the dissolved solids are believed to be a significant factor, a separate analysis may be made on several samples by separating the colloids by flocculation, then evaporating the remaining water to determine a weight for total dissolved solids. Then subtract that amount from the weight of total solids in all samples taken in that area. The resulting concentrations are expressed in grams/m^3. An alternative method that eliminates the need to subtract dissolved solids is filtering the sediment through a glass or cellulose filter. This requires less time than evaporation.

The volume of sample needed varies with the amount of suspended sediment. Small amounts of sediment in the water require larger samples to obtain reliable data. If sediment concentration is greater than 100 grams/m^3, 1 liter of sample is sufficient. For concentrations less than 20 grams/m^3, 10 liters are necessary. Because most sampling containers are 1 liter or less, multiple samples may be needed in some sites.

As mentioned earlier, two designs allow for instantaneous sampling at any chosen depth (point sampler) or continuous sampling over a range of depths (depth-integrating sampler). The cost for sediment samplers ranges from US \$200 to US \$800, depending on design.

Because suspended sediment concentration varies significantly vertically, the sampler provides more complete information about sediment in the volume of water than does the Secchi disk. However, the information above makes it apparent that collecting suspended sediment samples requires significantly more time and expense.

The purpose of either approach in a remote sensing project is to calibrate image data with water-borne sediment. In most projects it is sufficient to complete a survey with a Secchi disk.

Depth Measurements

Water depths may be measured by a sounding rod, a weighted cord or cable, or an ultrasonic sounding device. For water up to 5 m deep a graduated rod may be the fastest and most convenient. If the bottom material is soft, a flat plate on the foot of the rod will prevent inserting the rod into muck. A graduated cord or cable is needed for deeper water. A flat-shaped weight will keep the cord vertical and prevent the weight from sinking into a soft bottom.

The ultrasonic sounder is the fastest method. An ultrasonic signal is emitted through the water, and the device records the travel time of the signal to the bottom and back to the instrument. The display on most recent models is a digital reading of depth. Handheld models no bigger than a small flashlight are available. The depth range for such an instrument is from 0.6 m to 79 m, with an accuracy of ±0.1%. The operator immerses the instrument in the water a few centimeters, pointed vertically, and reads a result in seconds. Another depth sounder device is available for simultaneously measuring depth and temperature at intervals of 5 feet from surface to bottom. It is simply dropped into the water on a cord and then retrieved with data stored and ready for recording.

SPECTRAL RESPONSE OF SNOW

Snow has a very distinctive spectral reflectance. In the visible and near infrared up to about 800 nm snow reflects nearly 100%, essentially the same as clouds. This is strong enough to have a saturating effect on the sensors, causing them to read maximum values. At wavelengths longer than 800 nm, snow reflectance diminishes rapidly to almost zero at 1500 nm and stays near that value. Clouds, with nonselective scatter, continue reflecting at nearly 100% throughout the infrared (Figure 8.3). Furthermore, wet snow tends to have a lower reflectance than dry snow. Hence, some assessment of snow condition is possible. Melting near the edges of a snow area increases the free water content of the snow. This makes the snow edges darker, and the total snow

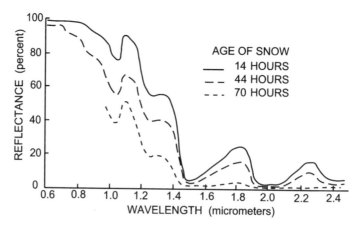

FIGURE 8.3. Change of snow reflectance with aging. Adapted from O'Brien and Munis (1975).

area appears smaller in wavelengths greater than 1500 nm, compared to the area of snow in the visible bands. The difference in area provides an assessment of the amount of melting.

DESCRIPTION OF SNOW COVER

Terminology for classification and description of snow cover in the field provides a uniform basis for taking notes and mapping the vertical profiles of snow cover. An International Snow Classification provides descriptive terms and symbols for snow properties, including crystal size, cohesion, and wetness. Five degrees of classification for each of these properties is based on magnitude or state of snow metamorphism. Further description in the classification includes depth, density hardness, and temperature for vertical profiles of the snow cover. An abbreviated version of the International Snow Classification, shown in Figure 8.4, provides a rapid method for field description (U.S. Army Corps of Engineers, 1952). Instruments are available for more precise and quantitative information on these variables, but this descriptive approach would serve the need for describing snow conditions in most remote sensing projects.

MEASUREMENT OF SNOW COVER

Three variables are essential for the practical use of snow in hydrologic forecasting: depth of snow, areal extent of snow cover, and the water equivalent of snow cover. The area of snow cover may be measured on imagery and requires no field work, but depth and water equivalent will require field work unless a regularly maintained snow course is near the study area.

Snow Courses

Federal and state agencies in the United States maintain hundreds of snow courses over the years, and collect data on depth and water equivalence at regular intervals (Garstka, 1964). The data are used for estimating future water yields (total water) and probable rates of runoff in the annual snow melt season. The data are available to the public. In the western United States, data from all agencies, also some private operators, are compiled in one place, and a complete data set is maintained by the USDA Natural Resources Con-

GRAIN NATURE	Map Symbol
New snow (Original crystal forms such as stars, plates, prisms, needles, and graupel are recognizable.)	Fa
Old snow, granular, fine-grained (Mean diameter is less than approximately 2 mm – like table salt.)	Db
Old Snow, granular, coarse-grained (Mean diameter is larger than approximately 2 mm – like coarse sand.)	Dd
Depth hoar (cup-shaped crystals 3 to 10 mm diameter)	De
HARDNESS (Use gloves)	
Soft (Four fingers can be pushed into the snow without effort.)	Kb
Medium hard (One finger can be pushed into the snow without effort.)	Kc
Hard (A pencil can be pushed into the snow without effort.)	Kd
Very hard (A knife can be pushed into the snow without effort.)	Ke
WETNESS (Use gloves)	
Dry (Snowball cannot be made)	Wa
Moist (Snow does not obviously contain liquid water, but snowball can be made.)	Wc
Wet (Snow obviously contains liquid water)	Wd
Slushy (Water can be pressed out.)	We

Examples:	New snow, dry and medium hard	FaKcWa
	Fine-grained snow, moist and hard	DbKdWe

FIGURE 8.4. Simplified field classification of snow types. Adapted from U.S. Army Corps of Engineers (1952).

servation Service (formerly Soil Conservation Service). In the central and eastern United States, a central body does not maintain snow course data, and some inquiry among federal and state agencies is necessary to find the right data source in a particular region.

A snow course consists of a traverse of sampling points, usually 10 or more. In a mountainous region the sample points may be 50–100 feet apart, and each point is clearly marked to assure that the same points are measured at each visit. The location of a snow course is carefully selected for representativeness of the site. Although the site should be selected in winter, it should not be fully marked and established until summer so specific sites with fallen timber or boulders on the ground can be avoided. Obviously, these obstructions would give erroneous data when inserting a sample tube from the snow surface above. In the western United States one snow course for a 260-km² (100-square-mile) drainage basin is considered adequate, although one course for 1000 km² (400-square-mile) area is not unusual. Snow course data should accumulate for 10 years or more before reliable statistical stream flow forecasting can be done.

A temporary snow course to be used as ground control for a remote sensing project would be justified in locations where no nearby existing snow course is available. The purpose of a temporary snow course, like any ground data in remote sensing, is to calibrate imagery of a particular time period to the snow conditions of the same time period. For that application, long-term data may not be required.

Making a Snow Survey

Sample Tubes

In a pinch, any cylinder can be used to cut a sample of snow for measuring both depth and water equivalent. However, as the snow becomes deeper and ice layers may be present at depth, more specialized equipment is better. A device known as the federal snow sampler has cutting teeth at the base of a cylinder. This sampler has an inside diameter such that 1 ounce (oz) of weight is equivalent to 1 inch depth of water. It is easy to operate, lightweight to carry, and provides good accuracy. The field person needs several sections of the cylindrical sampling tubes, a driving wrench for pushing them into the snow, and a weighing scale with wire hangers for holding the snow-filled cylinder for weighing.

Snow Pillows

A fluid-filled circular pillow on the ground beneath the snow may be used to measure the weight of snow. Hydrostatic pressure in the pillow is measured by a float-level recorder and converted to weight. Data from the pillows may be observed in the field or telemetered by radio to a receiving station. Possibly the snow over a pillow can develop an ice layer that forms a bridge above the pillow. If that happens, additional snow produces no change in the recorded weight of overlying snow (Starosolszky, 1987).

Travel to a snow course usually can be done by a snowcat or snowmobile, though often field personnel must be prepared to ski or snowshoe part of the way. Some staff snow gauges are read at a distance using binoculars or from the air in a light airplane. The advantage in either case is that less time is spent in travel and more sites can be observed. The disadvantage to observing from a distance is that only depth can be observed and no weight is available, so water content cannot be computed unless a telemetered snow pillow has been installed.

9

Applying Concepts of Field Work to Urban Projects

Previous chapters are concerned with field work for remote sensing projects that are specifically focused on vegetation, soils, or water. The objectives of many projects, however, are concerned with surface features that combine these three basic materials. Probably the most common examples of this would be any study involving land use mapping or the analysis of urban areas. In such projects many map categories are composed of earth materials similar to soil or rock (e.g., building materials), along with vegetation or water in some proportion. This chapter considers the issues and problems associated with field work discussed in previous chapters and applies them to an urban project.

PROJECT OBJECTIVES

The most common objective for urban remote sensing projects over the years has been to produce a map of urban land use. This type of map is always in demand as a practical tool for planners and others who need to study urban structure. Preparation of land use maps may also lead to other applications such as studying urban change over time by comparing land use maps made several years apart.

Other applications of remote sensing to urban studies have been less oriented to producing a map than finding relationships among image features and urban socioeconomic variables, for example, house density measured on an image and income derived from census data for a city. These projects may lead to models for estimating population, income, or poverty distribution.

Urban hydrology has been another useful application of remote sensing in cities. Runoff is greatly altered by the addition of extensive impervious areas during urbanization. Remote sensing has contributed to these studies by providing a means to determine the location and extent of impervious surfaces.

This chapter considers approaches to field work for these three objectives, urban land use maps, urban socioeconomic analysis, and urban hydrology.

SPECTRAL RESPONSE IN URBAN AREAS

The high frequency of spatial change in urban areas, along with the great mixture of surface materials within each map category, produces distinctive urban spectral responses, making most of the urban categories separable in imagery. Yet, urban spectral responses can also be said to resemble familiar curves from vegetation, soil and rock, or water.

Relatively pure vegetation pixels occur in few places in cities, for example, parks or cemeteries. Vegetation plays a decreasing role in other urban land use categories from low-density residential areas with much vegetated surface to others such as high-density commercial or industrial areas with almost no vegetation. The influence of vegetation on the spectral response curve is therefore similar to those obtained in natural areas that make a transition from high-density vegetation through intermediate stages to bare ground similar to that shown in Figure 6.4. The spectral responses of many urban construction materials are similar to soil responses with the absence of water absorption bands. Concrete and asphalt, for example, may occur on opposite ends of a soil line in a plot of infrared data against red data, with asphalt appearing among dark soil pixels in a diagram such as Figure 7.1.

AERIAL PHOTOGRAPHS

One useful advantage to working in urban areas is that many of the individual features of interest, such as houses, can be seen very well on aerial photographs. This makes it possible to utilize photographs for actual measurement or counting of features to a great extent in urban areas. With photographs land use categories can be identified accurately; homogeneous areas can be delimited visually; buildings, vehicles, and trees can be counted; streets and driveways can be measured. Much may remain to be done in the field, but photographs greatly reduce the amount of actual field measurements in cities.

URBAN LAND USE

Sampling

The principles of sampling do not change in urban areas. The object, as always, is to obtain a sample that is representative of each class. The criteria for homogeneity must be defined for each cover type. The category "low density residential," for example, might be defined as a single-family residential area having vegetation of more than 60%. The exact threshold of vegetative cover for each category would depend on the project objectives, and would likely be different in other locations. The guide for minimum area of a sample site, outlined in Chapter 2, applies to urban areas. The difficulty is that sufficiently homogeneous areas of adequate size for sampling are often more difficult to find for each category in cities.

Field Operations

Because many measurements of urban features can be made on air photos, field work for urban land use mapping often consists of verifying photos and land use categories. Field checking aerial photographs is always important with any project. With features so dynamic as in a city, field checks are essential even for photos of recent date. Cluster maps of satellite data then may be compared with aerial photographs to identify the appropriate category for each cluster.

URBAN SOCIOECONOMIC STUDIES

Objectives

Early studies of urban areas using remote sensing showed relationships between such variables as house density and poverty (Mumbower & Donoghue, 1967; McCoy & Metivier, 1973; Henderson & Utano, 1975). The relationships were determined by correlation analysis between selected features measured on photographs and socioeconomic data obtained from corresponding census blocks. Sometimes no maps were produced.

Field Operations

If satellite imagery is used for socioeconomic studies of urban areas, much of the actual calibration may still rely on air photos. House density may be measured on high-resolution satellite images, but indicators of house quality and characteristics of neighborhoods are best identified on air photos in combination with trips to the field. In U.S. urban areas a decrease in vegetation with increasing single-family house density provides a change in spectral response that will distinguish areas. However, air photos often must be used to determine the actual number of houses per unit of area, or the amount of vegetation.

Much of the distinction among housing areas of various economic levels lies in certain features that may not be evaluated on air photos. This includes such things as the amount of on-street parking, the presence of litter and trash, and the amount and quality of landscaping, building materials, or architectural styles (Hadfield, 1963; Lindgren, 1971). Obtaining information on these details requires field work in each of the neighborhoods of interest. Much of the needed information can be found simply by driving through the areas and tabulating observations for each of the variables on prepared forms that note street and block with categories for marking observations.

URBAN HYDROLOGY

Objectives

An important use of remote sensing in urban hydrology is to determine cover types and their influence on storm runoff. Methods devised for estimating storm runoff assume that water hitting the surface during a rainfall event must either infiltrate, run off, or be held as surface retention. Evaporation, another process for disposing of rainfall, is usually not considered in such estimates. A focus of interest, therefore, is the degree of imperviousness to water that each cover type has. The more impervious a surface, the less there will be infiltration and the greater the storm runoff from that surface.

Two approaches are often applied to estimating urban runoff. One approach is some variation on the so-called rational formula that estimates flow based on rainfall input, area, and a runoff coefficient. The runoff coefficient is the fractional component of rainfall that becomes runoff. The American Society of Civil Engineers (1969) devised a table of runoff coefficients for various urban land uses. Concrete streets, for example, have a runoff coefficients of 0.80–0.95. Single-family residential areas have coefficients of 0.30–0.50. Suburban residential, meaning less dense areas, have coefficients of 0.25–0.40. Parks and cemeteries run as low as 0.10 and up to 0.25 for runoff coefficients.

A second approach is called the Soil Conservation Service (SCS) method (U.S. Soil Conservation Service, 1975). Although the name of the agency has changed, the old name for the method has remained. In this method a set of curve numbers was established for urban land use categories. A curve number is similar to the runoff coefficient, but varies with the type of underlying soil and the amount of antecedent rainfall on the surface. For example, a curve number for parks and cemeteries might range from 39, where infiltration is highest, to 80, for soils that are much less permeable. In addition, the SCS tables relate urban residential runoff to a combination of lot size and the percentage of impervious area on the lot.

The concepts of runoff coefficients and runoff curve numbers are applied to remote sensing with good results because a current land use map can be produced with known areas for each category. By applying either a curve number or a runoff coefficient to each category, or to each pixel, a land use map becomes a runoff map. Some of the researchers using these techniques are Ragan and Jackson (1980), Fan (1991), and Al-Ghamdi (1991). Ridd (1995) devised a method for dealing with the complex mixture of surfaces in urban areas by classifying a surface based on the proportions of vegetation, impervious surface, and soil at each point.

Field Operations

Field work for a project involving urban hydrology will involve verifying land uses on aerial photographs. In addition notations on construction materials such as road surfaces and roofing may be needed. Vegetation density estimates may be needed also. If the SCS method is being used, the soil map for the area should be verified for the hydrologic soil group at each location. If no soil information exists, field personnel may need to make infiltration tests in the field. In this way relative (high, medium, low) infiltration information similar to the soil hydrologic groups can be obtained.

APPENDIX 1

Selected Bibliography on Field Methods and Related Topics Not Cited in the References

GENERAL

Alford, M., Tuley, P., Hailstone, E., & Hailstone, J. (1974). The measurement and mapping of land resource data by point sampling on aerial photographs. In E. C. Barrett & L. F. Curtis (Eds.), *Environmental remote sensing* (pp. 113–126). London: Edward Arnold.

Benson, A. S., Draeger, W. C., & Pettinger, L. R. (1971). Ground data collection and use. *Photogrammetric Engineering, 37*, 1159–1167.

Bonn, F. (1976). Some problems and solutions related to ground truth measurements for thermal infra-red remote sensing. *Photogrammetric Engineering, 42*, 1–11.

Buckland, S., & Elston, D. (1994). Use of ground truth to correct land cover estimates from remotely sensed data. *International Journal of Remote Sensing, 15*, 1273–1282.

Campbell, J. B., & Browder, J. O. (1995). Field data for remote sensing analysis: SPOT data, Rondonia, Brazil. *International Journal of Remote Sensing, 15*, 333–350.

Clark, R. N., & Roush, T. L. (1984). Reflectance spectroscopy: Quantitative analysis techniques for remote sensing applications. *Journal of Geophysical Research, 89*(B7), 6392–6340.

Congalton, R. G. (1991). Review of assessing the accuracy of classifications of remotely sensed data. *Remote Sensing of Environment, 37*, 35–46.

Curran, P. J., & Williamson, H. D. (1985). The accuracy of ground data used in remote sensing investigations. *International Journal of Remote Sensing, 6*, 1637–1651.

Dozier, J., & Strahler, A. H. (1983). Ground investigations in support of remote sensing. In R. N. Colwell (Ed.), *Manual of remote sensing* (2nd ed., pp. 959–986). Falls Church, VA: American Society of Photogrammetry.

Harris, R. (1987). Ground data collection. In R. Harris (Ed.), *Satellite remote sensing: An introduction* (pp. 35–48). New York: Routledge & Kegan Paul.

Hay, A. M. (1979). Sampling designs to test land-use accuracy. *Photogrammetric Engineering and Remote Sensing, 45*, 529–533.

Hotchkiss, N. J. (1995). *A comprehensive guide to land navigation with GPS* (2nd ed.). Herndon, VA: Alexis.

Hutchinson, C. F. (1982). Techniques for combining Landsat and ancillary data for digital classification improvement. *Photogrammetric Engineering and Remote Sensing, 48*, 123–130.

Jackson, R., Clarke, T., & Moran, S. (1992).Bidirectional calibration results for 11 Spectralon and 16 BaSO4 reference reflectance panels. *Remote Sensing of Environment, 40*, 231–239.

Lee, K. (1975). Ground investigations in support of remote sensing. In R. G. Reeves (Ed.), *Manual of remote sensing* (1st ed., pp. 805–856). Falls Church, VA: American Society of Photogrammetry.

Letham, L. (1998). *GPS made easy: Using global positioning systems in the outdoors* (2nd ed.). Seattle, WA: The Mountaineers.

Lintz, J., Brennan, P. A., & Chapman, P. E. (1976). Ground-truth and mission operations. In J. Lintz & D. S. Simonett (Eds.), *Remote sensing of environment* (pp. 412–437). Reading, MA: Addison-Wesley.

Mitchell, C. W. (1973). *Terrain Evaluation*. London: Longman.

Philipson, W. R. (Ed.). (1996). *Photographic interpretation*. Bethesda, MD: American Society of Photogrammetry and Remote Sensing.

Smedes, H. W. (1975). The truth about ground truth. In *Proceedings of the 10th International Symposium on Remote Sensing of the Environment* (pp. 821–823). Ann Arbor, MI: Environmental Research Institute of Michigan.

Steven, M. D. (1987). Ground truth: An underview. *International Journal of Remote Sensing, 8*, 1033–1038.

Townshend, J. R. G. (Ed.). (1981). *Terrain analysis and remote sensing*. London: Allen & Unwin.

Verbyla, D. L. (1995). *Satellite remote sensing of natural resources*. Boca Raton, FL: Lewis.

VEGETATION

Andrews, M. H., Noble, I. R., & Lange, R. T. (1979). A non-destructive method for estimating the weight of forage on shrubs. *Australian Rangeland Journal, 3*, 225–231.

Andrews, M. H., Noble, I. R., Lange, R. T., & Johnson, A. W. (1981). The measurement of shrub forage weight: Three methods compared. *Australian Rangeland Journal, 3*, 74–82.

Avery, T. E. (1967). *Forest measurements*. New York: McGraw-Hill.

Beatly, J. C. (1969). Biomass of desert winter annual plant populations in southern Nevada. *Oikos, 20*, 261–273.

Bentley, J. R., Seegrist, D. W. , & Blakeman, D. A. (1970*). A technique for sampling low shrub vegetation by crown volume classes* (Forest Service Research Note PSW-215, U. S. Department of Agriculture). Washington, DC: U.S. Government Printing Office.

Biging, G., Congalton, R., Murphy, E. (1991). A comparison of photo interpretation and ground measurements of forest structure. In *Proceedings of the 56th Annual Meeting of the American Society of Photogrammetry and Remote Sensing* (pp. 6–15). Falls Church, VA: American Society of Photogrammetry.

Bourdo, E. A. (1956). A review of the General Land Office Survey and its use in quantitative studies of former forest. *Ecology, 37,* 754–768.

Bouriaud, O., Soudani, K., & Bréda, N. (2003). Leaf area index from litter collection: Impact of specific leaf area variability within a beech stand. *Canadian Journal of Remote Sensing, 29,* 371–380.

Brown, J. K. (1976). Estimating shrub biomass from basal stem diameters. *Canadian Journal of Forest Research, 6,* 153–158.

Canfield, R .H. (1942). Measurement of grazing use by the line interception method. *Journal of Forestry, 42,* 192–194.

Cook, C. W. (1973). Forage utilization, daily intake, and nutrient value of desert range. In D. N. Hyder (Ed.), *Arid shrublands: Proceedings of the Third Workshop of the United States/Australian Rangeland Panel* (pp. 47–50). Tucson, AZ: Society of Range Management.

Davis, J., Tueller, P. T., & Bruner, A. D. (1972). Estimating forage production from shrub ring widths in Hot Creek Valley, Nevada. *Journal of Range Management, 25,* 398–402.

Dean, S. I., Burkhardt, W., & Meeuwig, R. (1981). Estimating twig and foliage biomass of sagebrush, bitterbrush, and rabbitbrush in the Great Basin. *Journal of Range Management, 34,* 224–227.

Eklundh, L., Hall, K., Eriksson, H., Ardö, J., & Pilesjö, P. (2003). Investigating the use of Landsat thematic mapper data for estimation of forest leaf area index in southern Sweden. *Canadian Journal of Remote Sensing, 29,* 349–362.

Evans, R. A., & Jones, M. B. (1958). Plant height times ground cover versus clipped samples for estimating forage production. *Agronomy Journal, 50,* 504–566.

Gerbermann, A. M., Cuellas, J. A., & Wiegand, C. L. (1976). Ground cover estimated from aerial photographs. *Photogrammetric Engineering, 42,* 551–556.

Gobena, A. (1984). *Influence of sagebrush (Artemesia tridentata) invasion on crested wheatgrass and its detection by Landsat imagery.* Unpublished thesis, Utah State University, Logan.

Grigal, D. F., & Ohman, L. F. (1977). *Biomass estimation for some shrubs from northeastern Minnesota* (Forest Service Research Note NC-226, North

Central Forest Experiment Station). St. Paul, MN: U.S. Department of Agriculture.

Hall, R. J., Davidson, D. P., & Peddle, D. R. (2003). Ground and remote estimation of leaf area index in Rocky Mountain forest stands, Kananaskis, Alberta. *Canadian Journal of Remote Sensing, 29*, 411–427.

Harlan, J. C., Haas, R. H., Boyd, W. E., & Deering, D. W. (1979). Determination of range biomass using Landsat. In *Proceedings of the 13th International Symposium on Remote Sensing of the Environment* (pp. 659–673). Ann Arbor, MI: Environmental Research Institute of Michigan.

Heylingers, P. C. (1968). Quantification of vegetation structure on vertical aerial photographs. In G. A. Stewart (Ed.), *Land Evaluation* (pp. 251–262). Melbourne: Macmillan.

Hilmon, J. B. (1959). Determination of herbage weight by double-sampling: Weight estimate and actual weight. In *Proceedings, Techniques and Methods of Measuring Understory Vegetation* (pp. 20–25). Tifton, GA: U.S. Department of Agriculture, Forest Service.

Holben, B. N., Tucker, C. J., & Fan, C. J. (1980). Spectral assessment of soybean leaf area and leaf biomass. *Photogrammetric Engineering and Remote Sensing, 46*, 651–656.

Jackson, R. D., Pinter, P. J., Idso, S. B., & Reginato, R. J. (1979). Wheat spectral reflectance: Interaction between crop configuration sun elevation and azimuth angle. *Applied Optics, 18*, 3730–3732.

Jensen, J. R. (1983). Biophysical remote sensing. *Annals of the Association of American Geographers, 73*, 111–132.

Johnson, G. R. (1976). *Remote estimation of herbaceous biomass*. Unpublished thesis, Colorado State University, Fort Collins, CO.

Jordan, C. F. G. (1969). Derivation of Leaf Area Index from the quality of light on the forest floor. *Ecology, 50*, 663–666.

Kaufmann, M. R. (1981). Automatic determination of conductance, transpiration and environmental conditions in forest trees. *Forest Science, 27*, 817–827.

Kaufmann, M. R., Edminster, C. B., & Troendle, C. A. (1982). Leaf area determination for subalpine tree species in the central Rocky Mountains (Research Paper RM-238, Rocky Mountain Forest and Range Experiment Station). Fort Collins, CO: U.S. Forest Service, U.S. Department of Agriculture.

Knipling, E. B. (1970). Physical and physiological basis for the reflectance of visible and near infrared radiation from vegetation. *Remote Sensing of the Environment, 1*, 155–159.

Küchler, A. W. (1967). *Vegetation mapping*. New York: Ronald Press.

Ludwig, J. A., Reynolds, J. F., & Whiston, P. D. (1975). Size-biomass relationships of several Chiuahuan Desert shrubs. *American Midland Naturalist, 94*, 451–461.

Marks, B. (1981). *A method of estimating canopy cover density over large areas*. Unpublished thesis). University of California, Santa Barbara.

Mueller-Dumbois, D., & Ellenburg, H. (1974). *Aims and methods of vegetation ecology*. New York: Wiley.

Murray, R. B., & Jacobson, M. Q. (1982). An evaluation of dimension analysis for predicting shrub biomass. *Journal of Range Management, 34*, 451–454.

Parker, K. W. (1951). *A method for measuring trend in range condition in National Forest ranges*. U.S. Department of Agriculture, Forest Service, Washington, D.C.

Pearson, R. L., Tucker, C. J., & Miller, L. D. (1976). Spectral mapping of shortgrass prairie biomass. *Photogrammetric Engineering and Remote Sensing, 42*, 317–323.

Phillips, D. R., & Saucier, J. R. (1981). Cruising procedures for estimating total stand biomass (Georgia Forest Research Paper, No. 14). Atlanta, GA: Georgia Forestry Commission.

Pieper, R. D. (1978). *Measurement techniques for herbaceous and shrubby vegetation*. Department of Animal and Range Science, New Mexico State University, Las Cruces, NM.

Pontailler, J., Hymus, G. J., & Drake, B. G. (2003). Estimation of leaf area index using ground-based remote sensed NDVI measurements: Validation and comparison with two indirect techniques. *Canadian Journal of Remote Sensing, 29*, 381–387.

Poulton, C. E. (1972). A comprehensive remote sensing legend system for the ecological characterization and annotation of natural and altered landscapes. In *Proceedings of the 8th International Symposium on Remote Sensing of the Environment* (pp. 393–408). Ann Arbor, MI: Environmental Research Institute of Michigan.

Rautiainen, M., Stenberg, P., Nilson, T., Kuusk, A., & Smolander, H. (2003). Application of a forest reflectance model in estimating leaf area index of Scots pine stands using Landsat-7 ETM reflectance data. *Canadian Journal of Remote Sensing, 29*, 314–323.

Sader, S. A. (1987). Forest biomass, canopy structure, and species composition relationships with multipolarization L-band synthetic aperature radar data. *Photogrammetric Engineering and Remote Sensing, 53*, 193–202.

Seed, E. D., & King, D. J. (2003). Shadow brightness and shadow fraction relations with effective leaf area index: Importance of canopy closure and view angle in mixedwood boreal forest. *Canadian Journal of Remote Sensing, 29*, 324–335.

Tucker, C. J. (1977). Spectral estimation of grass canopy variables. *Remote Sensing of Environment, 6*, 11–26.

Tucker, C. J., Elgin, H. H., & McMurtrey, J. E. (1979). *Relationship of red and photographic infrared spectral radiance's to alfalfa biomass, forage water content, percentage canopy cover, and severity of drought stress* (NASA

Technical Memorandum 80272). Greenbelt, MD: Goddard Space Flight Center.

U.S. Department of Agriculture, Forest Service. (1966). *Range analysis and management training guide*. Lakewood, CO: U.S. Department of Agriculture, Forest Service, Rocky Mountain Region.

Went, I. E., Thompson, R. C. A., & Arson, K. N. (1975). *Land systems inventory: Boise National Forest, Idaho, A basic inventory for planning and management*. Ogden, UT: U.S. Forest Service Intermountain Region.

With, J. R. (1967). The sampling unit and its effect on salt bush yield estimates. *Journal of Range Management, 20*, 323–325.

AGRICULTURE

Asrar, G., Kanemasu, E. T., & Yoshida, M. (1985). Estimates of leaf-area index from spectral reflectance of wheat under different cultural practices and solar angles. *Remote Sensing of Environment, 17*, 1–11.

Cinemas, E. T. A. (1974). Seasonal canopy reflectance patterns of wheat, sorghum and soybean. *Remote Sensing of the Environment, 3*, 43–47.

Cinemas, E. T. A., Heinemann, J. C. L., Bagel, C. O., & Powers, W. L. (1977). Using Landsat data to estimate evapotranspiration of winter wheat. *Environmental Management, 1*, 515–520.

Kumar, M., & Monteith, J. L. (1982). Remote sensing of crop growth. In H. Smith (Ed.), *Plants and the Daylight Spectrum*. New York: Academic Press.

Richardson, A. J. (1975). Plant, soil and shadow reflectance components of row crops. *Photogrammetric Engineering, 41*, 1401–1407.

Steven, M. D., Briscoe, P. V., & Jaggard, K. W. (1983). Estimation of sugar beet productivity from reflection in the red and near infrared spectral bands. *International Journal of Remote Sensing, 4*, 325–334.

GEOLOGY AND SOILS

Abu, P. A., & Nizeyimana, E. (1991). Comparisons between spectral mapping units derived from SPOT image texture and field soil maps. *Photogrammetric Engineering and Remote Sensing, 57*, 397–405.

Belcher, D. J. (1948). Determinations of soil conditions from aerial photographs. *Photogrammetric Engineering, 14*, 482–488.

Buckingham, W. F., & Somber, S. O. (1983). The characterization of rock surfaces formed by hydrothermal alteration and weathering: Application to remote sensing. *Economic Geology, 78*, 664–674.

Gerbermann, A. H., & Neher, D. D. (1979). Reflectance of varying mixtures of a clay soil and sand. *Photogrammetric Engineering and Remote Sensing, 45*, 1145–1151.

Hodgson, J. M. (1978). *Soil Sampling and Soil Description*. Oxford, UK: Oxford University Press.

Hunt, G. R., & Salisbury, J. W. (1970). Visible and near-infrared spectra of minerals and rocks. I: Silicate minerals. *Modern Geology, 1*, 238–300.

Hunt, G. R., Salisbury, J. W., & Lenhoff, C. J. (1971a). Visible and near-infrared spectra of minerals and rocks: III. Oxides and hydroxides. *Modern Geology, 2*, 195–205.

Hunt, G. R., Salisbury, J. W., & Lenhoff, C. J. (1971b). Visible and near-infrared spectra of minerals and rocks: IV. Sulphides and sulphates. *Modern Geology, 3*, 1–14.

Hunt, G. R., Salisbury, J. W., & Lenhoff, C. J. (1973a). Visible and near-infrared spectra of minerals and rocks: VII. Acidic igneous rocks. *Modern Geology, 4*, 217–224.

Hunt, G. R., Salisbury, J. W., & Lenhoff, C. J. (1973b). Visible and near-infrared spectra of minerals and rocks: VI. Additional silicates. *Modern Geology, 4*, 85–106.

Hunt, G. R., Salisbury, J. W., & Lenhoff, C. J. (1974) Visible and near-infrared spectra of minerals and rocks: IX. Oxides and hydroxides. *Modern Geology, 5*, 23–38.

Hunt, G. R., & Ashley, R. P. (1979). Spectra of altered rocks in the visible and near infrared, *Economic Geology, 74*, 1613–1629.

Mitchell, J. E., West, N. E., & Miller, I. E. (1966). Soil physical properties in relation to plant community patterns in the shadscale zone of northwestern Utah. *Ecology, 47*, 627–630.

Moore, G. K., & Waltz, F. A. (1983). Objective procedures for lineament enhancement and extraction. *Photogrammetric Engineering and Remote Sensing, 49*, 641–647.

Pieters, C. M., & Englert, P. A. J. (Eds.). (1993). *Remote Geochemical Analysis: Elemental and Mineralogical Composition*. Cambridge, UK: Cambridge University Press.

Ray, R. G. (1960). *Aerial photographs in geologic interpretation and mapping* (U.S. Geological Survey Professional Paper 373). Washington, DC: U.S. Government Printing Office.

Siegal, E. R., & Goetz, A. F. H. (1977). Effect of vegetation on rock and soil type discrimination. *Photogrammetric Engineering and Remote Sensing, 43*, 191–196.

Soil Survey Staff. (1951). *Soil Survey Manual* (U.S. Department of Agriculture Handbook No. 18). Washington, DC: U.S. Government Printing Office.

Stoner, E. R., Baumgardner, M. F., & Weismiller, R. C. A., Biehl, L. L., & Robinson, B. F. (1980). Extension of laboratory-measured spectra to field conditions. *Soil Science of America Journal, 44*, 572–574.

Wise, D. U. (1982). Linesmanship and the practice of linear geo-art. *Geological Society of America Bulletin, 93*, 886–888.

Wright, J. S., Vogel, T. C., Pearson, A. R., & Messmore, J. A. (1981). *Terrain*

Analysis Procedural Guide for Soil (ETL-0254). Ft. Belvoir, VA: U.S. Army Corps of Engineers, Engineering Topographic Laboratories.

LAND USE

Anderson, J. R., Hardy, E. E., Roach, J. T., & Witmer, R. E. (1976). *A Land Use and Land Cover Classification for Use with Remote Sensing Data* (U.S. Geological Survey Professional Paper 964). Washington, DC: U.S. Government Printing Office.

Bryand, C. R., LeDrew, E. F., Marois, C., & Cavayas, F. (Eds). (1989). *Remote Sensing and Methodologies of Land Use Change Analysis.* Waterloo, Canada: Department of Geography, University of Waterloo.

Campbell, J. B. (1996). Land use inventory. In W. R. Philipson (Ed.), *Photographic Interpretation* (Chapter 11). Bethesda, MD: American Society for Photogrammetry and Remote Sensing.

Downs, S. W., Sharma, G. C., & Bagwell, C. (1977). *A Procedure for a Ground Truth Study of a Land Use Map of North Alabama Generated from Landsat Data.* NASA Technical Note, NASA TN D-8420. Washington, DC: U.S. Government Printing Office.

Fitzpatrick, K. A. (1977). The strategy and methods for determining accuracy of small and intermediate scale land use and land cover maps. In *Proceedings of the 2nd Pecora Symposium* (pp. 339–361). Bethesda, MD: American Society of Photogrammetry.

URBAN

Forster, B. (1980). Urban control for Landsat data. *Photogrammetric Engineering and Remote Sensing, 46,* 539–545.

Forster, B. (1980). Urban residential ground cover using Landsat digital data. *Photogrammetric Engineering and Remote Sensing, 46,* 547–558.

Forster, B. (1983). Some urban measurements from Landsat data. *Photogrammetric Engineering and Remote Sensing, 49,* 1693–1707.

Lo, C. P., & Welch, R. (1980). Chinese urban population estimations. *Photogrammetric Engineering and Remote Sensing, 46,* 246–253.

McCoy, R. M., & Metivier, E. (1973). House density as a measure of socioeconomic conditions on aerial photos. *Photogrammetric Engineering, 39,* 43–47.

Mumbower, L., & Donoghue, J. (1967). Urban poverty study. *Photogrammetric Engineering, 33,* 610–618.

Olds, E. B. (1949). The city block as a unit for recording and analyzing urban data. *Journal of the American Statistical Association,* December, 485–500.

Rowntree, R. C. A. (Ed.). (1984). Ecology of the urban forest: Part I. Structure and composition. *Urban Ecology, 8,* 178–190.

Slonecker, E. T., Jennings, D. B., & Garofalo, D. (1994). Remote sensing of impervious surfaces: A review. *Remote Sensing Reviews, 20,* 227–255.

WATER

Chow, V. T., Maidment, D. R., & Mays, L. W. (1988). *Applied Hydrology.* New York: McGraw-Hill.

Davies-Colley, R. J., Vant, W. N., & Smith, D. G. (1993). *Colour and clarity of natural waters.* London: Ellis Horwood.

Hutchinson, G. E. (1957). *A treatise on limnology: Vol. 1. Geography, physics and chemistry.* New York: Wiley.

Khorram, S. (1980). Water quality mapping from Landsat digital data. *International Journal of Remote Sensing, 2,* 143–153.

Khorram, S., & Cheshire, H. M. (1985). Remote sensing of water quality in the Neuse River estuary, North Carolina. *Photogrammetric Engineering and Remote Sensing, 51,* 329–341.

Lathrop, R. G. (1992). Landsat Thematic Mapper monitoring of turbid inland water quality. *Photogrammetric Engineering and Remote Sensing, 58,* 465–470.

Liedtke, T., Roberts, A., & Luternauer, J. (1995). Practical remote sensing of suspended sediment concentration. *Photogrammetric Engineering and Remote Sensing, 61,* 167–175.

Lyzenga, D. R. (1981). Remote sensing of bottom reflectance and water attenuating parameters in shallow water using aircraft and Landsat data. *Journal of Remote Sensing, 2,* 71–82.

Owe, M., Brubaker, K., Ritchie, J., & Rango, A. (Eds.). (2000). *Remote Sensing and Hydrology 2000.* Wallingford, UK: International Association of Hydrological Sciences.

Rainwater, F. H., & Thatcher, L. L. (1960). *Methods for Collection and Analysis of Water Samples* (U.S. Geological Survey Water Supply Paper 1454). Washington, DC: U.S. Government Printing Office.

Slonecker, E. T., Jennings, D. B., & Garofalo, D. (1994). Remote sensing of impervious surfaces: A review. *Remote Sensing Reviews, 20,* 227–255.

Verdin, J. P. (1985). Monitoring water quality conditions in a large western reservoir with Landsat imagery. *Photogrammetric Engineering and Remote Sensing, 51,* 343–353.

Wetzel, R. G., & Likens, G. E. (2000). *Limnological Analysis* (3rd ed.). New York: Springer.

World Meteorological Organization. (1974). *Guide to Hydrological Practices* (3rd ed., WMO—No. 168). Geneva, Switzerland: Author.

World Meteorological Organization. (1992). *Snow Cover Measurements and Areal Assessment of Precipitation and Soil Moisture* (WMO—No. 749). Geneva, Switzerland: Author.

APPENDIX 2

FIELD NOTE FORMS

This compilation of forms for field notes should be regarded as suggested formats for note taking. They would best be used as guides to be adapted to the needs and specifications of a particular project. Forms may be combined to create new forms appropriate for special applications.

Note that the field forms contain certain categories of information which may be organized hierarchically. For example, the category "vegetation cover" may be subdivided as follows: "Physiognomy" (forest, shrub, grass), then each of those groups might be subdivided as "species composition," "cover density," "plant height/vigor." Each subdivision should have a definition stating, for example, the percentage of trees needed to identify an area as "forest", and the percentage of a species which must be present to name it a particular type of forest, along with the quantitative break points between high, medium and low cover density or crown closure.

Each sheet should also contain some information on the physical character of the site including topography (slope, aspect, relief), soil (texture, color, organic matter), hydrologic condition (poorly or well drained). Reference materials related to the site are also identified: map sheet, aerial photo ID, ground photo ID, UTM coordinates.

The project objectives will determine how far into the hierarchy it is necessary to proceed for each cover type. Some cover types might require identification of the first level only, and others would require intensive measurement at other levels in the hierarchy.

FIELD DATA
LINE TRANSECT FOR VEGETATION SURVEYS
(Based on BLM pace method)

Project Name _____   Date/Time _____/_____

Site ID/Photo Reference_____/_____   Observer _____

GPS Grid _____ Coordinates X: _____ Y: _____

TRANSECT DATA (100 points)

TRANSECT SUMMARY

Live Vegetation ____% Rock _____% Bare Soil ____% Litter ____%

Species	Percent	Comments:
1 _____	____%	
2 _____	____%	
3 _____	____%	
4 _____	____%	

Instruction: The five arrays of 3 × 20 provide space for three levels of cover data at each "hit" along a 100 point transect. Observations are made at the toe of each 5 ft to 6 ft stride (e.g., each right foot step). One hundred points make a 500 ft to 600 ft transect, which may be in a straight line or not, as fits the sample site. Alternatively, exact intervals along a measuring tape may be observed. The totals of each category equal the percent of that category covering the surface.

Adapted from Bureau of Land Management Form 7322.

FIELD DATA
CROPS AND PASTURES

Project Name _____   Date/Time _____/_____

Site ID/Photo Reference_____/_____   Observer _____

GPS Grid _____   Coordinates X: _____ Y: _____

GENERAL DESCRIPTION

Slope angle: _____ Slope aspect: _____

Cover type: Crop (name) _____ Pasture: planted _____, natural _____

Stubble _____ Bare (plowed) _____

Weeds only _____ Other _____

Planting technique: row _____, drilled _____, broadcast _____, other _____

Plant height: (nearest 0.5 feet) _____ Row width _____, Row direction _____

PLANT STATUS

Phenologic state: sprouting _____, flowering _____, mature _____,

senescent _____, other _____

Ground covered: _____ 0%–20% _____ 20%–40% _____ 40%–60%

_____ 60%–80% _____ 80%–100%

Recent cutting or mowing _____

Weed infestation: type _____, percent of area _____

Disease infestation: type _____, percent of area _____

Insect infestation: type _____, percent of area _____

SOIL

Soil series _____, Texture _____, Color _____

Surface moisture condition: Dry _____, Moist _____, Saturated _____

Root zone moisture: Dry _____, Moist _____, Saturated _____

COMMENTS

Adapted from Joyce (1978).

FIELD DATA
EXTRACTIVE LAND USES

Project Name _____   Date/Time _____/_____

Site ID/Photo Reference_____/_____   Observer _____

GPS Grid _____   Coordinates X: _____   Y: _____

ACTIVITY TYPE

_____ Sand Pit   _____ Clay

_____ Gravel Pit   _____ Chert

_____ Stone, dimensioned   _____ Lignite

_____ Stone, crushed   _____ Coal

_____ Limestone (for cement)   _____ Heavy metals
                                            (iron, copper, etc.)

Other _____

STATUS

_____ in use   _____ abandoned and barren

_____ abandoned and revegetated

WATER

Impounded water all year? _____ Occasional water? _____

COMMENTS

Adapted from Joyce (1978).

134

FIELD DATA
FOREST CHARACTERISTICS

Project Name _____   Date/Time _____/_____

Site ID/Photo Reference _____/_____   Observer _____

GPS Grid _____ Coordinates X: _____ Y: _____

GENERAL DESCRIPTION

Slope Angle _____ Slope Aspect _____

STAND DESCRIPTION

Species	DBH	Crown Width	Height	Phenologic Status	Percent Cover
1					
2					
3					

Note: DBH = Diameter of trunk at breast height; phenologic status of plants in each layer (flowering, leaf sprouts, dormant, etc.); numbers 1, 2, 3 refer to layers.

SOIL

Soil Series _____ Texture _____ Color _____

Soil Moisture: Dry _____ Moist _____ Saturated _____

COMMENTS

Instruction: Record species, phenologic status, and cover information for each forest layer at sample points. Use a new sheet for each sample point.

FIELD DATA
GENERAL PURPOSE

Project Name _____   Date/Time _____/_____

Site ID/Photo Reference _____/_____   Observer _____

GPS Grid _____   Coordinates X: _____   Y: _____

LAND USE _____

TOPOGRAPHIC INFORMATION

Slope angle _____ Slope Aspect _____

SOILS INFORMATION

Texture: sandy _____ silt _____ clay _____ loam _____

   stony _____ soil absent _____ parent material _____

Moisture: dry _____ moist _____ saturated _____

   signs of seasonally waterlogged soil _____

Color: _____

VEGETATION INFORMATION

Physiognomic Type _____

	Dominant Species	% Cover
Top Layer	1 _____	_____
	2 _____	_____
Intermediate	1 _____	_____
	2 _____	_____
Ground Layer	1 _____	_____
	2 _____	_____

COMMENTS

Adapted from Joyce (1978).

FIELD DATA
GPS STATION OBSTRUCTION DIAGRAM

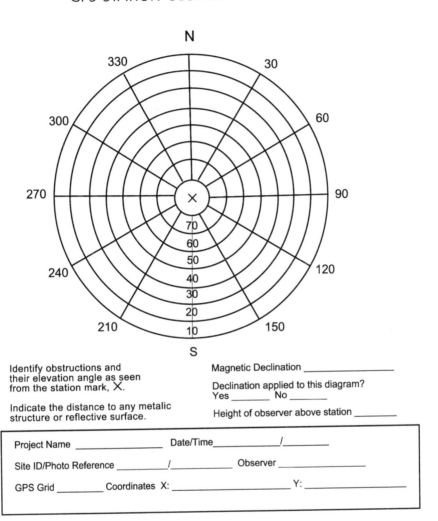

Identify obstructions and
their elevation angle as seen
from the station mark, X.

Indicate the distance to any metalic
structure or reflective surface.

Magnetic Declination _____

Declination applied to this diagram?
Yes _____ No _____

Height of observer above station _____

Project Name _____ Date/Time_____/_____

Site ID/Photo Reference _____/_____ Observer _____

GPS Grid _____ Coordinates X: _____ Y: _____

Adapted from Geomatics Canada (1995).

FIELD DATA
PHOTOGRAPHS

Project Name _____   Date/Time _____/_____

Site ID/Photo Reference _____/_____   Observer _____

GPS Grid _____   Coordinates X: _____   Y: _____

Position No.	Roll No.	Exposure No.	Camera Settings	Direction	Sky	Target Description

Adapted from Joyce (1978).

FIELD SPECTROMETER DATA

Project Name _____   Date/Time _____/_____

Site ID/Photo Reference _____/_____   Observer _____

GPS Grid _____   Coordinates X: _____ Y: _____

Instrument ID _____ Number of Scans Averaged _____

Geometry of observer's position relative to incident radiation: (sketch or description)

Height of instrument above ground _____ Height of instrument above target _____

Scan No.	Target	Time	Sky Condition	Comments

COMMENTS: (Sketches may be helpful)

Instruction: This form provides for scans from several positions at each target, or scans from adjacent targets of the same type at the same site (e.g., same species of shrub). Use a separate form for each new site. Comments on individual scans could mention if a new reference scan was used, or if a sudden gust of wind arose. Comments for the site should mention nearby features that may affect overall incident radiation on the target, such as nearby trees, buildings, or landform features.

FIELD DATA
URBAN AREAS

Project Name _____   Date/Time _____/_____

Site ID/Photo Reference _____/_____   Observer _____

GPS Grid _____   Coordinates X: _____   Y: _____

SITE CHARACTERISTICS

High Density Urban (   )*
   Cover Materials

           Predominantly Concrete (est.)   _____%

           Predominantly Asphalt   _____%

           Other Materials (roofing)   _____%

                  built-up   _____%

                  metal   _____%

Comments*** _____

Low Density Urban (   )**
   Cover Materials

      Inert Materials

           Concrete   _____%

           Asphalt   _____%

           Roofing Type/Color

               _____/_____   _____%

               _____/_____   _____%

      Vegetative

           Grass   _____%

           Deciduous Trees   _____%

           Evergreen Trees   _____%

           Mixed Trees   _____%

           Mixed Grass/Trees   _____%

Comments*** _____

*High density is an area essentially devoid of vegetation. Small scattered parcels (no larger than 100 ft in longest dimension) of vegetation may total as much as 10% of the area.

**Low density is an area with within which inert materials (roofs, concrete, asphalt) are predominant, but with up to 45% of the surface covered with vegetation, including overtopping trees, occurring in small, scattered parcels with the maimum dimension of each parcel no greater than 200 ft. Parks and cemeteries may be regarded as special cases exceeding the limits set above.

*** Appropriate comments include identification of land use, height of buildings (number of stories), roof pitch (flat, moderate, steep). Also, include any other information that may be pertinent to an overhead sensor.

Adapted from Joyce (1978).

140

FIELD DATA
WETLAND AREAS

Project Name _____   Date/Time _____/_____

Site ID/Photo Reference _____/_____   Observer _____

GPS Grid _____   Coordinates X: _____ Y: _____

VEGETATION TYPE

Pure stand (monotypic)—species: _____

Mixed with fewer than six vascular species—dominant species:

_____

Mixed with more than six vascular species—dominant species:

_____

Note: Define dominant as more than 5% of vegetation present.

HOMOGENEITY

Category	Distribution*	Density**
Vegetation pattern (continuous, zones, clumps) _____		
Barren areas _____		
Open water _____		
Sparse vegetation/barren _____		
Sparse vegetation/water _____		
Other (describe) _____		

*Distribution may be: E (evenly distributed), C (center only), P (peripheral only)
**Density may be: H (high, >90% cover), M (medium <70% cover), L (low, < 50%)

Plant size and status	Species	Height	% Living/Dead	Stage of Growth*
1. _____				
2. _____				
3. _____				
4. _____				

*Stage of Growth categories include: dormant, seedlings, immature, mature, flowering.

Adapted from Joyce (1978).

APPENDIX 3

Metadata Online Resources

Metadata is information about the content, accuracy, and description of geospatial data sets. It can be said that metadata is to a data set as a map legend is to a map. In this time of extensive exchange of geospatial data, it becomes increasingly important to know the original source, accuracy, location, and description of data before being transferred from site to site. The key purpose of metadata is to facilitate access to geospatial information.

At present, many different metadata formats exist, and a number of government and private entities are now concerned with establishing standards of format and quality for metadata. In this regard, the Federal Geographic Data Committee has taken a lead in helping set standards and disseminating information about them to the public. Their website, given below, is a rich source of information on standards, access, and availability of metadata.

Anyone involved with remote sensing imagery is by definition faced with large amounts of geospatial data. State and federal government agencies have many digital images and geospatial maps based on field work. The widespread dissemination and use of these maps has made their metadata descriptions increasingly important.

This appendix provides access to a selection of online information resources about metadata standards, development, and availability.

GENERAL RESOURCES FOR METADATA

- FGDC Metadata Website
  [Check this website for updates to this list of websites.]
  http://www.fgdc.gov/metadata/metadata.html

- Minnesota Land Management Information Center (LMIC)—Metadata Resources
  http://www.lmic.state.mn.us/chouse/meta_help.html

- NOAA Coastal Services Center—Wonderful World of Coastal Metadata
  http://www.csc.noaa.gov/metadata/

- USGS—Geology Division—Information and Tools for Formal Metadata
  http://geology.usgs.gov/tools/metadata/

- International Federation of Library Associations and Institutions
  http://www.ifla.org/II/metadata.htm#geo

- U.S. National Aeronautics and Space Agency (NASA). Directory Interchange Format (DIF) Writer's Guide
  http://gcmd.gsfc.nasa.gov/difguide/difman.html

GEOSPATIAL DATA CONVERTERS AND LOCATORS

- Chuck Taylor Freeware Toolbox—UTM to Lat/Lon
  http://home.hiwaay.net/~taylorc/toolbox/geography/geoutm.html

- U.S. Army Corps of Engineers—state plane, geographic, and UTM
  http://crunch.tec.army.mil/software/corpscon/corpscon.html

- Texas Bureau of Economic Geology—lat/lon, scale, length, area, time, radians
  http://www.beg.utexas.edu/GIS/tools/index.html

- Australian GeoDynamics Cooperative Research Center—WGS84
  http://www.agcrc.csiro.au/info/resources/proj.html

- U.S. Federal Communications Commission—decimal degrees and degrees/minutes/seconds
  http://www.fcc.gov/mb/audio/bickel/DDDMMSS-decimal.html

- National Geodetic Survey—NAD83 and NAD27
  http://www.ngs.noaa.gov/cgi-bin/nadcon.prl

- USGS National Mapping Information Geographic Names Information System—latitude and longitude for specific place names
  http://geonames.usgs.gov/

SAMPLE METADATA

- National Geospatial Data Clearinghouse (NSDI)
  http://clearinghouse1.fgdc.gov/

- Energy and Environmental Information Resource Center—NWRC/University of Louisiana—Lafayette
  http://eeirc.nwrc.gov/metadata_fgdc.htm

- Louisiana Geographic Information Center Data Catalog
  http://lagic.lsu.edu/datacatalog

- Mecklenburg County NC Metadata Server
  http://maps.co.mecklenburg.nc.us/metaweb/

METADATA CREATION SOFTWARE

- USGS—Metadata Editor—tkme (Windows)
  http://geology.usgs.gov/tools/metadata/tools/doc/tkme.html

- USGS—Metadata Editor—xtme (UNIX)
  http://geology.usgs.gov/tools/metadata/tools/doc/xtme.html

- Corpsmet 95 metadata creation freeware from USACE
  http://corpsgeo1.usace.army.mil

- NOAA—ArcView Metadata Collector
  http://www.csc.noaa.gov/metadata/text/download.html

- NOAA—MetaScribe Metadata Template Builder
  http://www.csc.noaa.gov/metadata/text/metascribe.htm

- USGS National Biological Resource Division—Metamaker
  http://www.umesc.usgs.gov/metamaker/nbiimker.html

- DataTracker (Commercial Metadata Software)
  http://www.theaxongroup.com/dtfgdc.html

- Spatial Metadata Manager Software (SMMS)
  http://www.intergraph.com/gis/smms/

- M3CAT—ISO-based Metadata Creation Freeware
  http://www.intelec.ca/technologie_a.html

KEYWORD THESAURI
AND ATTRIBUTE LABEL DEFINITION SOURCES

- Global Change Master Directory (GCMD)—Science Keywords and Associated Directory Keywords
  http://gcmd.gsfc.nasa.gov/Resources/valids

- Geographic Names Index Service (GNIS)
  http://geonames.usgs.gov/

- Cowardin USFWS Wetlands Classification System
  http://h2osparc.wq.ncsu.edu/info/wetlands/class.html

- NLCD Land Cover Class Definitions
  http://landcover.usgs.gov/classes.html

- Anderson Land Cover Classification System
  http://landcover.usgs.gov/pdf/anderson.pdf

- NBII Systematics
  http://www.nbii.gov/disciplines/systematics.html

- Glossary of Geological Terms
  http://www.geotech.org/survey/geotech/dictiona.html

- Glossary of Landform and Geologic Terms, official reference for soil survey applications, the National Soil Survey Center
  http://soils.usda.gov/classification/taxonomy/main.htm

GOVERNMENT INFORMATION LOCATOR SERVICE (GILS)

- U.S. Geological Survey. Government Information Locator Service.
  http://www.gils.net/

- U.S. National Archives and Records Administration (NARA). Guidelines for the Preparation of GILS Core Entries.
  http://www.ifla.org/documentslibraries/cataloging/metadata/naragils.txt
  http://www.ifla.org/documentslibraries/cataloging/metadata/bull95-3.txt

INTERNATIONAL INFORMATION LOCATOR SERVICE

- Australia. Information Management Steering Committee (IMSC). Architecture For Access To Government Information.
  http://www.defence.gov.au/imsc/imsctg/imsctg1a.htm

- Canada. Government Information Locator Service.
  http://gils.gc.ca

References

Adams, J. B., & Smith, M. O. (1986). Spectral mixture modeling: A new analysis of rock and soil types at the Lander I site. *Journal of Geophysical Research, 91*(B8), 8098–8112.

Al-Ghamdi, S. (1991). *Use of Landsat Thematic Mapper for the SCS runoff technique*. Unpublished dissertation, University of Utah, Salt Lake City.

American Society of Civil Engineers. (1969). *Design and Construction of Sanitary and Storm Sewers*. Manuals and Reports on Engineering Practices. No. 37.

Andrew, M. H., Noble, I. R., & Lange, R. T. (1979). A non-destructive method for estimating the weight of forage on shrubs. *Australian Range Journal, 1*, 225–231.

Andrew, M. H., Noble, I. R., Lange, R. T., & Johnson, A. W. (1981). The measurement of forage weight: Three methods compared. *Australian Range Journal, 3*, 74–82.

Bartolucci, L. A., Robinson, B. F., & Silva, L. F. (1977). Field measurements of the spectral response of natural waters. *Photogrammetric Engineering and Remote Sensing, 43*, 595–598.

Black, C. A. (1965). *Methods of soil analysis: Part I. Physical and mineralogical properties*. Madison, WI: American Society of Agronomy.

Bowman, R. A., Guenzi, W. D., & Savory, D. J. (1991). Spectroscopic method for estimation of soil organic matter. *Soil Science Society of America Journal, 55*, 563–566.

Bowers, S. A., & Hanks, R. J. (1965). Reflection of radiant energy from soil. *Soil Science, 100*, 130–138.

Breimer, R. F., van Kekem, A. J., & van Reuler, H. (1986). *Guidelines for Soil Survey and Land Evaluation in Ecological Research* (Man and the Biosphere (MAB) Technical Note #17). Paris, France: UNESCO.

Campbell, J. B. (1996). *Introduction to Remote Sensing* (2nd ed.). New York: Guilford Press.

Clark, R. N., Swayze, G. A., Gallagher, A. J., King, T. V. V., & Calvin, W. M. (1993*). The U.S. Geological Survey Digital Spectral Library: Version 1. 0.2 to 3.0 microns* (U.S. Geological Survey Open File Report 93-592). Washington, DC: U.S. Government Printing Office.

Congalton, R. G., & Green, K. (1999). *Assessing the Accuracy of Remotely Sensed Data: Principles and Practices.* Boca Raton, FL: Lewis.

Curran, P. J. (1983). Estimating Green LAI from multispectral photography. *Photogrammetric Engineering and Remote Sensing, 49*, 1709–1720.

Curran, P. J., Foody, G. M., Kondratyev, K. Y., Kozoderov, V. V., & Fedchenko, P. P. (1990). *Remote Sensing of Soil and Vegetation in the USSR.* London: Taylor & Francis.

Curtiss, B., & Goetz, A. (1994). Field Spectroscopy: Techniques and instrumentation. In *Proceedings of the International Symposium on Spectral Sensing Research* (pp. 195–203).

Daubenmire, R. F. (1959). Canopy coverage method of vegetation analysis. *Northwest Science, 33*, 43–64.

Daughtry, C. S. T., Vanderbilt, V. C., & Pollara, V. J. (1982). Variability of reflectance with sensor altitude and canopy type. *Agronomy Journal, 74*, 744–751.

Davies-Colley, R. J., Vant, W.N., & Smith, D. G. (1993). *Colour and clarity of natural waters.* London: Ellis Horwood.

Fan, Y. (1991). *Prediction of Runoff Using Landsat Thematic Mapper.* Unpublished thesis, University of Utah, Salt Lake City.

Fitzpatrick-Lins, K. (1981). Comparison of sampling procedures and data analysis for a land-use and land-cover map. *Photogrammetric Engineering and Remote Sensing, 47*, 349–366.

French, G. T. (1999). *Understanding the GPS.* Albany, NY: Delmar Learning.

Geomatics Canada, Geodetic Survey Division. (1995). *GPS Positioning Guide.* Ottawa: Natural Resources Canada.

Garstka, W. U. (1964). Snow and snow survey. In V. T. Chow (Ed.), *Handbook of Applied Hydrology* (pp. 10-1–10-57). New York: McGraw-Hill.

Gobena, A. (1984). *Influence of Sagebrush* (Artemesia tridentata) *Invasion on Crested Wheatgrass and Its Detection by Landsat Imagery.* Unpublished thesis, Utah State University, Logan.

Goel, N. S. (1988). Models of vegetation canopy reflectance and their use in estimation of biophysical parameters from reflectance data. *Remote Sensing Reviews, 4*, 1–212.

Hadfield, S. A. (1963). *Evaluation of Land Use and Dwelling Unit Data Derived from Aerial Photography.* Urban Research Section, Chicago Area Transportation Study, Chicago.

Henderson, F. M., & Utano, J. (1975). Assessing urban socio-economic conditions with conventional air photography. *Photogrammetria, 31*, 81–89.

Hoffer, R. M., & Johannsen, C. J. (1968). Ecological potentials in spectral signature analysis. *Agricultural Experiment Station Journal No. 3479.* Purdue University. Also appears in P. L. Johnson (Ed.). (1969) *Remote sensing in ecology* (pp. 1–16). Athens: University of Georgia Press.

Hunt, G. R. (1977). Spectral signatures of particulate minerals in the visible and near infrared. *Geophysics, 42,* 501–513.

Hunt, G. R. (1979). Near-infrared (1.3–2.4 μm) spectra of alteration minerals—potential for use in remote sensing. *Geophysics, 44,* 1,974–1,986.

Hurn, J. (1989). *GPS: A Guide to the Next Utility.* Sunnyvale, CA: Trimble Navigation.

Jackson, R., Clarke, T., & Moran, S. (1992). Bidirectional direction results for 11 Spectralon and 16 BaSO4 reference reflectance panels. *Remote Sensing of Environment, 40,* 231–239.

Jensen, J. R. (1996). *Introductory Digital Image Processing: A Remote Sensing Perspective* (2nd ed.). Upper Saddle River, NJ : Prentice-Hall.

Jordan, C. F. (1969). Derivation of leaf-area index from quality of light on the forest floor. *Ecology, 50,* 663–666.

Joyce, A. T. (1978). *Procedures for Gathering Ground Truth Information for a Supervised Approach to a Computer-Implemented Land Cover Classification of Landsat-Acquired Multispectral Scanner Data* (NASA Reference Publication 1015). Washington, DC: U.S. Government Printing Office.

Justice, C. O., & Townshend, J. R. G. (1981). Integrating ground data with remote sensing. In J. R. G. Townshend (Ed.), *Terrain Analysis and Remote Sensing* (pp. 38–58). London: Allen & Unwin.

Kauth, R. J., & Thomas, G. S. (1976). The tasseled cap—A graphic description of the spectral-temporal development of agricultural crops as seen by Landsat. In *LARS: Proceedings of the symposium on machine processing of remotely sensed date* (pp. 4B-41–4B-51). West Lafayette, IN: Purdue University.

Lindgren, D. T. (1971). Dwelling unit estimation with color-IR photos. *Photogrammetric Engineering, 37,* 373–377.

Lord, D., Desjardins, R. L., & Dube, P. A. (1985). Influence of wind on crop canopy. *Remote Sensing of Environment, 18,* 113–123.

McCoy, R. M., Blake, J. G., & Andrews, K. L. (2001). Detecting hydrocarbon microseepage. *Oil and Gas Journal,* May 28, 40–45.

McCoy, R. M., & Metivier, E. D. (1973) House density versus socio-economic conditions. *Photogrammetric Engineering and Remote Sensing, 39,* 43–49.

Miller, L. D., & Pearson, R. L. (1971). Areal mapping program of the IBP grassland biome: Remote sensing of the productivity of the shortgrass prairie as input into biosystem models. In *Proceedings of the Seventh International Symposium on Remote Sensing of Environment* (pp. 165–205). Ann Arbor, MI: Environmental Research Institute of Michigan.

Milton, E. J. (1987). Principles of field spectroscopy. *International Journal of Remote Sensing, 8*(12), 1807–1827.

Mumbower, L. E., & Donoghue, J. (1967). Urban poverty study. *Photogrammetric Engineering, 33*, 610–618.

Munsell Products. (1973). *Munsell Soil Color Charts*. Baltimore, MD: Macbeth Color & Photometry Division of Kollmorgen Corporation.

Myers, V. I. (1970). Soil, water, and plant relations. In *Remote Sensing with Special Reference to Agriculture and Forestry* (pp. 253–297). Washington, DC: National Academy of Sciences.

Nicodemus, F. E. (1977). *Geometrical Considerations and Nomenclature for Reflectance* (National Bureau of Standards Monograph 160). Washington, DC: U.S. Government Printing Office.

O'Brien, H. W., & Munis, R. H. 1975. *Red and Near Infrared Reflectance of Snow* (U.S. Army Cold Regions Research and Engineering Laboratory (USACRREL) Research Report 332). Hanover, NH.

Parker, K. W. (1951). *A Method for Measuring Trend in Range Condition in National Forest Ranges*. Washington, DC: U.S. Department of Agriculture, Forest Service.

Poreda, S. F. (1992). *Vegetation Recovery and Dynamics Following the Wasatch Mountain Fire (1990), Midway, Utah*. Unpublished thesis, University of Utah, Salt Lake City.

Ragan, R. M., & Jackson, T. J. (1980). Runoff synthesis using urban hydrologic models. *Journal of the Hydraulics Division, American Society of Civil Engineers, 106*, 512–521.

Richardson, A. J., & Wiegand, C. L. (1977). Distinguishing vegetation from soil background. *Photogrammetric Engineering and Remote Sensing, 43*, 1541–1552.

Ridd, M. K. (1995). Exploring a V-I-S (vegetation–impervious surface–soil) model for urban ecosystems analysis through remote sensing: Comparative anatomy for cities. *International Journal of Remote Sensing, 1*, 100–111.

Robinson, B. F., & Biehl, L. L. (1979). *Calibration Procedures for Measurement of Reflectance Factor in Remote Sensing Field Research.*. Society of Photo-Optical Instrumentation Engineers, *196*, 16–26.

Salisbury, J. W. (1998). *Spectral Measurements Field Guide*. Washington, DC: Defense Intelligence Agency, Central Measurement and Signature Intelligence Office.

Silk, J. (1979). *Statistical Concepts in Geography*. London: Allen & Unwin.

Sohn, Y., & Rebello, N. S. (2002). Supervised and unsupervised spectral angle classifiers. *Photogrammetric Engineering and Remote Sensing, 68*, 1271–1280.

Specht, M. R., Needler, R. D., & Fritz, N. L. (1973). New color film for water-photography penetration. *Photogrammetric Engineering and Remote Sensing, 39*, 359–369.

Starosolszky, O. (1987). Instruments and methods for observation and methods. In O. Starosolszky (Ed.), *Applied Surface Hydrology* (pp. 177–297). Littleton, CO: Water Resources Publications.

Stohlgren, T. J., Bull, K. A., & Otsuki, Y. (1998). Comparison of rangeland vegetation sampling techniques in the central grasslands. *Journal of Range Management, 51,* 164–172.

Stohlgren, T. J., Falkner, M. B., & Schell, L. D. (1995). A Modified-Whittaker nested vegetation sampling method, *Vegetatio, 117,* 113–121.

Stoner, E. R., Baumgardner, M. F. (1981). Characteristic variations in reflectance of surface soils. *Soil Science Society of America Journal, 45,* 1161–1165.

Swain, P. H., & Davis, S. M. (Eds.). (1978). *Remote Sensing: The Quantitative Approach.* New York: McGraw-Hill.

U.S. Army Corps of Engineers. (1952). *Simplified Field Classification of Natural Snow Type for Engineering Purposes.* Corps of Engineers, Snow, Ice and Permafrost Establishment. Washington, DC: U.S. Government Printing Office.

U.S. Department of Agriculture, Forest Service. (1966). *Range Analysis and Management Training Guide.* Lakewood, CO: U.S. Department of Agriculture, Forest Service, Rocky Mountain Region.

U.S. Soil Conservation Service. (1975). *Urban Hydrology for Small Watersheds* (Technical Release No. 55). Washington, DC: Author.

Van Sickle, J. (2001). *GPS for land surveyors* (2nd ed.). Boca Raton, FL: CRC Press.

Verschuur, G. L. (1997). Transparency measurements in Garner Lake, Tennessee: The relationship between Secchi depth and solar altitude and a suggestion for normalization of Secchi depth data. *Lake and Reservoir Management, 13,* 142–153.

Williams, R. B. G. (1984). *Introduction to Statistics.* London: Macmillan.

Index

About the Author

Roger M. McCoy earned a BS degree in petroleum geology from the University of Oklahoma, and worked for an oil company for several years before starting graduate school. He obtained a master's degree in geography from the University of Colorado, followed by a PhD in geography with an emphasis in remote sensing at the University of Kansas. After short periods of teaching at the University of Illinois at Chicago and the University of Kentucky, Dr. McCoy taught at the University of Utah until his retirement in 1998. During that time he taught remote sensing and physical geography and conducted research in remote sensing of vegetation, soils, and hydrocarbons. He lives near Tucson with his wife, Sue, and continues his interests in research and writing.